Studies in Computational Intelligence

Volume 746

Series editor

Janusz Kacprzyk, Polish Academy of Sciences, Warsaw, Poland
e-mail: kacprzyk@ibspan.waw.pl

The series "Studies in Computational Intelligence" (SCI) publishes new developments and advances in the various areas of computational intelligence—quickly and with a high quality. The intent is to cover the theory, applications, and design methods of computational intelligence, as embedded in the fields of engineering, computer science, physics and life sciences, as well as the methodologies behind them. The series contains monographs, lecture notes and edited volumes in computational intelligence spanning the areas of neural networks, connectionist systems, genetic algorithms, evolutionary computation, artificial intelligence, cellular automata, self-organizing systems, soft computing, fuzzy systems, and hybrid intelligent systems. Of particular value to both the contributors and the readership are the short publication timeframe and the world-wide distribution, which enable both wide and rapid dissemination of research output.

More information about this series at http://www.springer.com/series/7092

Cristian Lai · Alessandro Giuliani
Giovanni Semeraro

Editors

Emerging Ideas on Information Filtering and Retrieval

DART 2013: Revised and Invited Papers

 Springer

Editors
Cristian Lai
Center for Advanced Studies, Research
and Development in Sardinia (CRS4),
Parco Scientifico e Tecnologico della
Sardegna
Pula
Italy

Giovanni Semeraro
Department of Computer Science
University of Bari
Bari
Italy

Alessandro Giuliani
Department of Electrical and Electronic
Engineering
University of Cagliari
Cagliari
Italy

ISSN 1860-949X   ISSN 1860-9503   (electronic)
Studies in Computational Intelligence
ISBN 978-3-319-88594-0   ISBN 978-3-319-68392-8   (eBook)
https://doi.org/10.1007/978-3-319-68392-8

Printed on acid-free paper

This Springer imprint is published by Springer Nature
The registered company is Springer International Publishing AG
The registered company address is: Gewerbestrasse 11, 6330 Cham, Switzerland

Foreword

This volume focuses on new challenges and emerging ideas in distributed information filtering and retrieval. It collects invited chapters and extended research contributions from DART 2013 (the 7th International Workshop on Information Filtering and Retrieval), held in Turin (Italy), on December 6, 2013 and co-located with the XIII International Conference of the Italian Association for Artificial Intelligence.

The main focus of DART was to discuss and compare suitable novel solutions based on intelligent techniques and applied to real-world applications. The chapters of this book present a comprehensive review of related work and state-of-the-art techniques. The authors, both practitioners and researchers, shared their results in several topics such as data leak protection on text comparison, natural language processing, ambient intelligence, information retrieval and web portals, knowledge management. Contributions have been carefully reviewed by experts in the area, who also gave useful suggestions to improve the quality of the volume.

Pula, Italy   Cristian Lai
Cagliari, Italy   Alessandro Giuliani
Bari, Italy   Giovanni Semeraro

Preface

With the increasing amount of available data, it becomes more and more important to have effective methods to manage data and retrieve information. Data processing, especially in the era of Social Media, is changing the users' behaviors. Users are ever more interested in information rather than in mere raw data. Considering that the large amount of accessible data sources is growing, novel systems providing effective means of searching and retrieving information are required. Therefore, the fundamental goal is making information exploitable by humans and machines.

This book collects invited chapters and extended research contributions from DART 2013 (the 7th International Workshop on Information Filtering and Retrieval), held in Turin (Italy), on December 6, 2013 and co-located with the XIII International Conference of the Italian Association for Artificial Intelligence.

The book is focused on researching and studying new challenges in intelligent information filtering and retrieval.

In Chapter "On the Use of Matrix Bloom Filters in Data Leak Protection," by Sergey Butakov, modifications to Matrix Bloom filters for data leak prevention systems are proposed. Bloom filters are fast tools for massive document comparison. The approach proposed in this chapter aims at improving density in matrix Bloom filters with the help of a special index to track documents uploaded into the system. Density is increased by combining a few documents in one line of the matrix to reduce the filter size and to address the problem of document removal. Special attention is paid to the negative impact of filter-to-filter comparison in matrix filters. A theoretical evaluation of the threshold for false positive results is described. The performed experiment outlines advantages and applicability of the proposed approach.

Chapter "A Hybrid Approach for the Automatic Extraction of Causal Relations from Text," by Antonio Sorgente, Giuseppe Vettigli, and Francesco Mele, proposes a hybrid approach for the discovery of causal relations from open domain text in English. The approach first identifies a set of plausible cause-effect pairs through a set of logical rules based on dependencies between words, then it uses Bayesian inference to filter and reduce the number of pairs produced by ambiguous patterns. The proposed model has been evaluated in the context of the SemEval-2010 task 8 dataset challenge. The results confirm the effectiveness of the rules for relation extraction and the improvements made by the filtering process.

Chapter "Ambient Intelligence by ATML Rules in BackHome," by Eloi Casals, José Alejandro Cordero, Stefan Dauwalder, Juan Manuel Fernández, Marc Solá, Eloisa Vargiu, and Felip Miralles, focuses on an implementation of state-of-the-art ambient intelligence techniques implemented in the telemonitoring and home support system developed by BackHome. BackHome is a EU project that aims to study the transition from the hospital to the home for people with disabilities. The proposed solution adopts a rule-based approach to handle efficiently the system adaptation, personalization, alarm triggering, and control over the environment. Rules are written in a suitable formal language, defined within the project, and can be automatically generated by the system taking into account dynamic context conditions and user habits and preferences. Moreover, rules can be manually defined by therapists and caregivers to address specific needs. The chapter also presents the concept of "virtual device", a software element that performs a mash-up of information coming from two or more sensors in order to make some inference and provide new information.

In Chapter "Dense Semantic Graph and Its Application in Single Document Summarisation," by Monika Joshi, Hui Wang, and Sally McClean, a novel graph generation approach is proposed. The proposal is an extension of an existing semantic graph generation approach. Semantic graph representation of text is an important part of natural language processing applications such as text summarisation. In particular, two ways of constructing the semantic graph of a document from dependency parsing of its sentences are studied. The first graph is derived from the subject–object–verb representation of sentence, and the second graph is derived from considering more dependency relations in the sentence by a shortest distance dependency path calculation, resulting in a dense semantic graph. Experiments show that dense semantic graphs perform better in semantic graph-based unsupervised extractive text summarization.

In Chapter "Web Architecture of a Web Portal for Reliability Diagnosis of Bus Regularity," by Benedetto Barabino, Cristian Lai, Roberto Demontis, Sara Mozzoni, Carlino Casari, Antonio Pintus, and Proto Tilocca, a methodology to evaluate bus regularity is proposed. In high-frequency transit services, bus regu- larity—i.e., the headway adherence between buses at bus stops—can be used as an indication of service quality in terms of reliability, by both users and transit agencies. The web architecture is the entry point of a Decision Support System (DSS), and offers an environment designed for experts in transport domain. The environment is composed of tools developed to automatically handle Automatic Vehicle Location (AVL) raw data for measuring the Level of Service (LoS) of bus regularity at each bus stop and time interval of a transit bus route. The results are represented within easy-to-read control dashboards consisting of tables, charts, and maps, able to perform fast AVL processing and easy accessibility in order to reduce the workload of transit operators. These outcomes show the importance of well-handled and presented AVL data, in order to use them more effectively, improving past analysis done by using, if any, manual methods.

Chapter "An Approach to Knowledge Formalization", by Filippo Eros Pani, Maria Ilaria Lunesu, Giulio Concas, and Gavina Baralla, is focused on the study of a process to formalize knowledge through an iterative approach mixing the top-down analysis for the structure of the knowledge of the specific domain and the bottom-up analysis for the information in the object of the specific domain. Just a few years after its birth, knowledge management drew the attention of the academic world, becoming a matter of intense study as the scientific community has always felt the need of making its members cooperate through the exchange and the reciprocal fruition of information. Many disciplines develop standardized formalization of knowledge, which domain experts can use to share information in the form of reusable knowledge. The proposed case study analyzes the knowledge on the domain of descriptions and reviews of Italian wines which can be found on the Internet: starting from an analysis of the information on the web, a taxonomy able to represent knowledge is defined.

Pula, Italy   Cristian Lai
Cagliari, Italy   Alessandro Giuliani
Bari, Italy   Giovanni Semeraro
October 2014

Contents

On the Use of Matrix Bloom Filters in Data Leak Protection

Sergey Butakov

Abstract Data leak prevention systems become a must-have component of enterprise information security. To minimize the communication delay, these systems
require quick mechanisms for massive document comparison. Bloom filters have
been proven to be a fast tool for membership checkup. Taking into account specific needs of fast text comparison this chapter proposes modifications to the Matrix
Bloom filters. Approach proposed in this chapter allows improving density in Matrix
Bloom filters with the help of special index to track documents uploaded into the
system. Density is improved by combining a few documents in one line of the matrix
to reduce the filter size and to address the problem of document removal. Special
attention is paid to the negative impact of filter-to-filter comparison in matrix filters. Theoretical evaluation of the threshold for false positive results is provided.
The experiment provided in the chapter outlines advantages and applicability of the
proposed approach.

Keywords Data leak protection · Bloom filters · Text search

1 Introduction

Data leak protection (DLP) systems become a must-have part of enterprise information security infrastructure. One of the main functions for these systems is content
filtering. Content filtering in DLP is different from the one in antivirus or spam protection applications. In DLP it relies on the internal sources for comparison rather than
on malware signature database maintained by the antivirus vendor. Simple content
filtering in DLP could be based on keyword screening. More advanced mechanisms
allow comparing text in question with a set of documents that are listed for internal purposes only. It is assumed that the comparison must be done on the level of
paragraphs or sentences and thus leads to the task of similar text detection. The task

S. Butakov (✉)
Information Security and Assurance Department, Concordia University of Edmonton, Edmonton,
AB, Canada
e-mail: sergey.butakov@concordia.ab.ca

© Springer International Publishing AG 2018
C. Lai et al. (eds.), *Emerging Ideas on Information Filtering and Retrieval*,
Studies in Computational Intelligence 746, https://doi.org/10.1007/978-3-319-68392-8_1

of fast detection of near-duplicate documents is not new. It has been studied for years, starting with applications in simple text search in the 1970s [1]. Additional momentum has been added later by massive projects in genome sets comparison and various research fields in internet study. The latter includes near—duplicate web page detection [2], filtering of web addresses [3], internet plagiarism detection [4], and many others. Although there are many methods for fast text comparison many of them are not suitable for large scale tasks. One of the approaches called Bloom Filters offer speed of comparison at the cost of generating some false positive results.

Bloom Filters (BFs) were introduced in 1970 as a tool to reduce the space requirements for the task of fast membership checkup [5]. As Bloom mentioned in the chapter, this method allows significant reduction in the memory requirements at the price of small probability of false positive results [5]. He also indicated that some applications are tolerable to false positives as they require two-stage comparison: the first step is to quickly identify if an object is a member of a predefined set and the second step is to verify the membership. Text comparison can be considered as one of such tasks when in the first phase the system must reduce the number of documents to be compared from hundreds of thousands to thousands. This step reduces the workload for the second phase of comparison where the system needs to find and mark similar segments in the documents.

1.1 Motivating Applications

Insider misuse of information is considered to be the top reason for data leaks [6]. Regardless of the fact that such misuses can be either intentional or unintentional, it is always helpful to have additional defense mechanisms embedded into the information flow to limit the impact of data leaks [7]. Commercial DLP systems are developed by major players on information security market such as Symantec or McAfee. One of the tasks for such systems is to make sure that documents belonging to the restricted group do not leave the network perimeter of an organization without explicit permission from an authorized person [8]. DLP systems can work on different layers. On the application layer they can analyze keywords, tags or the entire message [9]. The decision to allow the document to be transmitted outside should be made quickly and should be prone to false negative errors. False positive could be acceptable up to the certain level if there are embedded controls in the system for manual verification and approval. An additional desired feature for DLP systems is the ability to detect obfuscated texts where some parts of the text are not confidential but other parts are taken from the restricted documents. BFs could play a pre-selector role that would facilitate fast search by picking from the large archive of protected documents some limited number of candidate sources that could be similar to the document in question. A detailed comparison can be done using other methods that could be slower but more granular.

1.2 Contribution

BFs have been studied in details in the last four decades. In this chapter we reviewed the problems related to their practical implementation. In its original form, the BF can identify that the object is similar to one or more objects in a predefined set of objects but it cannot identify which one it is similar to. In other words BFs do not contain localization information. Matrix Bloom Filter (MBF) [10, 11] allows such identification but, as shown in the next section, may impose significant requirements on memory to maintain the set of documents. Dynamic BFs [12] allow better memory utilization and deletion of elements, but still lack localization property. Approach proposed in this chapter allows density improvement for an MBF by constructing special index to track documents uploaded to MBF. In addition to index, the document removal algorithm with localized workload has been proposed.

Experiment in the last section shows that compacting documents in the MBF can significantly reduce memory consumption by the filter at the cost of maintaining smaller index of the documents that have been uploaded into the filter.

2 Previous Works

The first part of this section discusses original BF and some major modifications that occurred to improve its performance and usability; the second part highlights the areas where BFs can be improved in applications related to near-similar document detection.

2.1 Bloom Filters with Variations

BF was introduced as memory-efficient method for the task of fast membership checkup [5]. The main idea of BF is to use randomized hash functions to translate an object from set $S = \{x_0, x_1, \ldots, x_{(n-1)}\}$ into a binary vector of fixed size m. BF employs k independent hash functions $h_0, h_1, \ldots, h_{k-1}$ to map each element to a random number over the range $\{0, \ldots, m - 1\}$. To insert element x into BF, the corresponding bits of BF—$\{h_i(x), 0 \leq i \leq k-1\}$—have to be set to 1. To check if element x' is already in BF the corresponding values of $\{h_i(x'), 0 \leq i \leq k-1\}$ must be calculated and compared with the values in the filter. If all corresponding bits are set in 1 than x' already exists in the filter. Due to randomized nature of h_i BF may produce a false positive [5]. The probability of false positive can be estimated as follows [13]:

$$p' \approx (1-^{kn/m})^k \tag{1}$$

As it can be seen from (1) the false positive probability can be reduced by increasing the filter size m. The optimal size of BF can be defined using n—number of elements in S and acceptable risk of false positive—p'. BF has distinctive features:

- BF has constant insert and checkup time due to the fixed size of the filter—constant m. This feature makes it an attractive tool to check membership in large collections because checkup time does not grow along with the collection size.
- Removal of any element from BF requires the entire filter to be recreated. The fact that a single bit in the filter can be set to 1 by many elements leads to the necessity to recreate the entire filter if one element has to be removed.
- If the number of elements in set {S} grows above n then the new value for m has to be calculated and filter must be recreated.

The last four decades have brought many variations to BFs, aiming to overcome problems of member elements removal, filter fixed size limitations, and some others. Counter BFs [14] use bit counters instead of a single bit value in the filter. This approach allows to store the number of times the particular index has been initialized. The element addition/removal to/from the filter can be done by increasing/decreasing corresponding counter. Although such an approach keeps the membership checkup operation simple and computationally effective, it increases memory requirements. For example, one byte counter that can store up to 255 initializations of the particular index would increase the BF memory consumption by the factor of 8.

Dynamic or matrix Boom filters (MBF) [12] handle the growing size of {S} by adding additional zero-initialized vectors of size m to the filter when number of elements in S reaches n. Each new vector is considered a brand new zero-initialized BF. Such an approach allows indefinite growth of S at the cost of additional checkups because the membership has to be checked in each row in the matrix. The number of checkups required can be considered as linear to the size of S and it is of $\theta(|\{S\}|/n)$ order. Removal of any text would cause the corresponding row to be removed from the matrix.

For the applied problems of document comparison, the speed of comparison can be considered as a priority metric, while the speed of update operations such as addition and deletion of elements is less of the priority. This metric selection is based on the assumption that a set of documents for the comparison is relatively steady if compared to the information flow that has to go through the filter.

2.2   Related Problems

Document comparison may be done with different level of granularity. One extreme approach used by *diff* utility on *NIX provides character level comparison. More relaxed approached compare texts in general in order to find near-to-duplicates on the document level. For example, competitions on plagiarism detection usually try to provide evaluation corpuses that would include obfuscated or paraphrased text [18].

More granular approaches usually work on the level of a few consecutive words or a string sequence of the same length without selecting word from the test. Such granularity allows to locate meaningful similarities in the texts while staying prone to minor text alternations. The sequence of words or characters can be called a shingle or a grammar [15, 16]. A full set of these sequences defines text $T = \{t_0, t_1, \ldots, t_{z-1}\}$. If sequencing is done on the word level, then z is close to the number of words in the document. If sequencing is done on the character level it could lead to excessive representations. Many algorithms with proven performance and robustness have been developed to address the issue of excessive representation. For example, Winnowing algorithm provides $d = 2/(w + 1)$ density of the selection where w is number of characters in the sequence (w-gram) [12].

If we are to compare two documents, T_i and T_j of length z_i and z_j respectively, and the comparison has to be done using BF, then m must satisfy the following inequality: $m \leq max\{z_i, z_j\}$. Obviously, if BF is to be used to compare one document T' with many documents $\{T\}$ then BF has to be large enough to fit all of these many documents. Straightforward sequencing of all the documents followed by placing shingles into the BF will not work because original BF is not local. In other words, if two documents contain exactly the same shingle (sequences of words or characters) and both documents were placed in BF, the same shingle flips to 1 same bits and therefore there will be no way to identify the source document. It is obvious that similar problem appears when there is a need to remove a document from the comparison set $\{T\}$. As Geravand and Ahmadi suggested, the localization problem can be addressed by matrix BF. They proposed to use a set of BFs where each BF represents shingles from one document [11]. The document that is scheduled for the checkup generates its own BF which will be compared with every row in matrix BF on the next step. Although a set of independent BFs does not look like a typical BF, it has one important property: the number of *XOR* operations to check the membership in the entire matrix BF is linear to the matrix size. This important performance property is achieved by using the same fixed m for all rows of matrix BF.

This approach has the disadvantage of inefficient memory consumption by each row in the matrix because of two factors: the requirement to have one document per row and the requirement to have constant m. The last one provides the possibility to calculate BF for T' only once and therefore dramatically improve the comparison speed. The first requirement can be relaxed with the additional indexing added to the matrix. The following section proposes such an index and outlines related algorithms.

3 Pseudo-Matrix Bloom Filter

Placing all the shingles for the entire document collection in one large BF is not an option in DLP. These systems require not only the confirmation that a particular shingle from the document in question belongs to the collection but also entail the need of linking the same shingles to the potential source document. The proposed data structure resembles the one proposed by Geravand & Ahmadi for the task of plagiarism detection. Pseudo-Matrix BF (PMBF) will consist of an unlimited number of BFs of the same size [11]. The major alternation is to use one BF to store more than one document where it is possible. From the Eq. (1) we can find m that would allow to limit p for a maximum of n elements [13]:

$$m \geq -\frac{n ln(p)}{(ln(2))^2} \qquad (2)$$

Keeping in mind that m is fixed for the entire PMBF to achieve the fast comparison speed, we can state that if there is a restriction to place one and only one document in single BF we can meet the following cases:

- Number of shingles in a document exceeds n, thus the entire document cannot be placed in a single row;
- Number of shingles in a document is much less than n, thus the BF is under-populated which makes it memory inefficient.

Suboptimal memory allocation would require p to be slightly less than n. To provide the such memory allocation by PMBF, two additional indexes have to be maintained: one to keep track of the remaining capacity of each row and another one to store indexes of documents placed in each row of PMBF. Figure 1 provides an example of a PMBF. The example assumes that each row in BLOOM FILTERS matrix is populated with two documents *(T)*. The example can be extended to any number of documents in a single row. INDEX part of the data structure links filter rows with uploaded texts *{T_i}*. For each row the corresponding INDEX list identifies the documents that are stored in this row. For example, Fig. 1 assumes that each document is stored in two rows. As the last part of the chapter shows in real applications there will be no such even distribution and many documents could be stored in one row. If BF' compiled from T' has a certain level of similarity with BF_i we can say that T' is probably similar to one of the documents *(T)* located in BF_i. The detailed comparison must be done to confirm this. It can be done with any appropriate method such as Levenstein's distance; this stage of comparison is out of scope of this chapter.

BLOOM FILTERS		INDEX
BF_0	----	
- - -	----	
BF_{h-2}: T_{998}, T_{999}	----	I_{998}; I_{999};
BF_{h-1}: T_{999}, T_{1000}	----	I_{999}; I_{1000};
BF_h: T_{1000}, T_{1001}	----	I_{1000}; I_{1001};
BF_{h+1}: T_{1001}, T_{1002}	----	I_{1001}; I_{1002};
- - -	----	
BF_{d-1}	----	

Fig. 1 Example of Pseudo-Matrix Bloom Filter with two documents in a row

The proposed data structure allows relatively easy document removal with no increase in memory consumption. If T_i to be removed, then only the row(s) that contain contains T_i has to be recreated. For example, in Fig. 1 if document T_{1000} are to be removed from the collection then only BF_{h-1} and BF_h have to be recreated making it relatively minor work from the computational perspective. Thus the additional work is not computationally expensive if comparing with matrix BF presented in [11] but there is higher shingle density in the matrix. The density of each individual filter in the matrix can be evaluated using the index. Under-populated filters can be filled up with additional text segments.

The proposed architecture has the following important features:

- Compared to an MBF proposed earlier, it increases density for each BF_i, therefore increasing its memory efficiency.
- Increased density leads to better computational efficiency of the data structure. Although the comparison computations are still of $\theta(n^*d)$ order but in this structure d is the number of rows required to allocate the document corpus instead of total number of documents. As the experiment below shows it could be few times less than the total number of documents.
- It keeps number of binary checkups linear to the size of the collection $\{T\}$.
- It allows relatively easy document removal.
- Matrix size is pseudo linear to the total number of shingles in the collection.

Probability of false positives is a known feature of BF—it provides great space advantage at the cost of known maximum probability of a false positive outcome of the comparison. To address this issue additional checkup has to be performed on all alarms triggered by a BF. Classical BFs do not provide false negative results but matrix BFs do. It happens because in case of the matrix BF one BF must be bit-wise compared with another one of the same size. If n is the maximum capacity of a BF than the number of elements that can be potentially inserted in the filter cannot exceed n for given maximum p'. In case of PMBF n is actually being substituted with n'—a maximum length of the document that can be inserted into the filter. Obviously n' is much less than n thus $n \sim= n'$ assumption may affect the quality of the search because the compared filters are overpopulated. For example, if we assume dictionary with $1e+4$ words and combination of three words to be one element then n—number of elements that can be potentially inserted in the filter is close to $1e+12$ thus making one filter to occupy few hundred gigabytes if acceptable level of false positives (p') is set to 5%. If overpopulated BFs are bit-wise compared they will be generating unacceptable level of false positives thus making them unusable. It means even if two filters carry some number of 1s in the same positions this has to be ignored up to the certain level e.g. some threshold must be used to avoid overwhelming number of false positives. The theoretical evaluation of this threshold is provided below.

Equation (1) is based on the probability to have k elements of the BF to be 1 after insertion of n independent elements. False positives will be produced by 1 s in the result of an AND operation on two filters. In this case the probability can be written as follows:

$$p' \approx (1 - e^{-kn/m})^2 \tag{3}$$

Equation (3) evaluates the probability for two bits and with the same index i in two independent BFs BF^a and BF^b of the same size m and same k hash functions to be turned into 1. The document similarity level proposed in [11] is calculated as follows:

$$S' = \frac{|N|'}{n} \tag{4}$$

where $|N|$ is number of 1s in the resulting filter. Since we know the probability for two bits to be set in 1 in two filters $|N|'$ can be evaluated as (5):

$$|N|' = p' * m \tag{5}$$

Therefore the final evaluation of the threshold looks should look like follows:

$$T' = \frac{\left(1 - e^{-kn/m}\right)^2 m}{n} \tag{6}$$

Equation (3) assumes that and have exactly the same number of elements inserted into both filters. If this is not the case then the threshold of BF^a false positive similarity to BF^b can be evaluated as follows:

$$T'_{ab} = \frac{\left(1 - e^{-kn/m_a}\right)\left(1 - e^{-kn/m_b}\right) m_a}{n} \tag{7}$$

It must be noted that evaluation (7) is asymmetric, e.g. $T'_{ab} \neq T'_{ba}$. Eq. (6) will be tested in the following section.

Example calculations of the size and density provided below show the memory efficiency of the proposed approach. Second part of the experiment shows applicability and limitations set by $n \sim= n'$ assumption in the proposed approach.

4   Experiment

Two experiments have been conducted to test the proposed approach of PMBF utilization. First experiment has been aimed to evaluate the compression ratio of the PMBF and second experiment has been conducted to study applicability of PMBF on real data.

4.1   Evaluation of Memory Consumption by PMBF

The goal of the first experiment was to show on the real data that PMBF has distinctive size advantage over matrix BF. In this case the experimental calculations have been performed on the simulated plagiarism corpus used for the PAN'2009 plagiarism detection competition [17]. The corpus contained 14,428 source documents simulating plagiarism. The average number of words in a document is about 38,000; 95% of all documents have less than 123,000 words. The calculations provided in the Table 1 were done with the following assumptions:

- Text shingling is done by sliding window on the word basis. Keeping in mind that shingles (consecutive phrases of y words) are relatively short, we can safely assume number of shingles to be equals to the number of words in the text. The number of words in a shingle may affect the granularity of the search in PMBF but this question is out of scope of this chapter.
- The order in which the documents are placed in the filter does not affect its memory capacity. In practice, this will not be absolutely true because if random documents are constantly being added and deleted to/from PMBF, it could lead to a fragmented index thus decreasing the performance of the index search. The issue can be addressed by setting up some threshold to protect under-populated rows from accepting short sequences of shingles. This option will be explored in the future research.

As it can be seen from Table 1, for PAN'09 corpus PMBF has compressed the information in 1:3.1 ratio comparing to MBF because the straightforward implementation of one text per row approach would require 14,428 rows in matrix instead of 4489. Moreover, in MBF about 5% of the documents would not fit a single row and therefore would be out of search scope. The option to increase n up to the size of the largest document in the corpus ($\sim$428,000 words) would increase an MBF size by the factor of 4.

Similar calculations performed for a snapshot of English Wikipedia are presented in the rightmost column of Table 1. The calculations have been performed with the same assumptions. The Wiki corpus included 3,917,453 documents. The compression ratio achieved by PMBF is slightly higher: $3,917,453/1,036,853 \sim= 3,78$. It can be explained by less even distribution of words in the corpus documents.

The fact that PMBF allows to store more than one document in the row also indicates that 95% of the documents will be collated in one or two rows of the matrix. The latter case covers those documents that start in one row and end in the next one; therefore, in 95% of cases only two rows will have to be recreated if such a document is scheduled for removal from the filter. On the related issue of the space reuse, it can be suggested that algorithms similar to garbage collection can be implemented to take care of the released space after a document has been removed from PMBF. The garbage collection can be implemented in the system's off-peak time when search results can be delayed.

Obviously, improvement in space allocation comes with a cost of more comparisons to be done to confirm the similarity of the document, if an MBF similarity of BF' and BF_i means that respective texts T' and T_i must be compared more closely. In the case of PMBF, the similarity of BF' and BF_i means that T' must be compared with more than one document. According to the example above, in 95% of cases on average T' will have to be compared with a minimum of 4 documents and a maxi-

Table 1 Filter size and operations for the corpus

Value	Explanation for PAN'09 corpus	English Wikipedia
$p = 0.05$	5% false positive probability	5%
$n = 123,000$	Filter of this size will accommodate 95% of texts. If text has more than 123,000 words it will be placed in more than one filter. If text has less than 123,000 words empty space will be occupied by another text	$n = 1,400$ (95%)
$m = 766,933$ bit.	See Eq. (2). 766,933 bit $\sim= 94$ KB	$m = 8730$ bit $\sim= 1.07$ KB
$d = 4489$	Total number of rows in PMBF = Total number of words in corpus/Number of words to fit in one row	$d = 1,036,853$
$M \sim= 411$ MB	Total size of PMBF to accommodate PMBF for the entire corpus	$M \sim= 8.43$ GB

mum of 8 documents because 95% of the documents are located in one or two rows. Therefore, on one hand increasing n helps to improve the compression, but it also increases the workload for the system component that performs detailed comparison.

4.2 Applicability of PMBF for DLP Tasks

Goal of the second experiment was to test the applicability of PMBF for text similarities detection. As it was mentioned earlier, DLP systems have to be prone to false negatives and keep false positives in reasonable range. In our case PMBF is used on the first stage of comparison which is to indicate the potential similarity among documents thus reducing the range of the detailed comparison from thousands of potential matches to more manageable number which can be processed on the second page by more precise but slower methods.

Two data sets were used for the evaluation. One data set included 500 documents with text inserts from other 50 source documents. Each randomly placed insert had been taken from only one of 50 sources. The length of the inserts varied from 5% to 50% of the total document length. Each insert was placed as one piece in the random place of the document. All the documents consisted of about 4000 words. Due to the fact that the documents in this corpus were about the same length the compression was not implemented in this part of the experiment and therefore each document had been placed in the separate row of PMBF. Each of 50 source documents has been compared with all 500 documents from this corpus.

The level of similarity was evaluated as number of ones in the same cells of the one row [11]. Due to Eqs. (6) and (7) it was naturally expected to see some similarity among totally different documents. We used Eq. (6) for evaluation because all the documents in this experiment had roughly the same size. The similarity level if calculated according to Eq. (6) for $n = 4000$, $k = 4$, and $m = 143700$ should be ∼0.399. The data on Fig. 2 shows the average background similarity level to be about 0.43 which is slightly above its theoretical evaluation provided by (6). The deviation can be explained by the fact of minor coincidental similarities in random documents caused by features of the natural language such as common general phrases, repeating words, etc.

The major question in this experiment was to see if the actual inserts from source documents produce higher level of similarity which is distinct from the background similarity level defined by Eqs. (6) or (7). The experiment indicated that for all 50 sources it was true. For each source document the documents that had some parts from that source produced distinctive pikes on the similarity level graph. An example of such graph for one of the sources is presented on Fig. 2. It shows 10 distinct similarity levels for 10 sources that included from 5% (leftmost pike) to 50% (rightmost pike) of the text from that particular source. Based on this part of the experiment we can state that PMBF are suitable for preliminary comparison of text documents in the cases where large portions—5% or above—were copied without variations from one of confidential documents.

Second phase of experiment was conducted to evaluate level of false positives and false negatives on the larger corpus with obfuscated text. Since MBF and PMBF should have same detection capacity the experiment was done on PMBF only due to limited resources. Training corpus from PAN'09 plagiarism detection competition [17] was used for PMBF performance evaluation. As it was mentioned earlier this part of experiment aimed at evaluation of $n \sim= n'$ assumption of the number of false negatives.

PAN'09 corpus consists of 14,428 documents [17]. Number of pair-wise comparisons for all the documents in the corpus would be about $(14*10^3*14*10^3)/2 \sim 10^8$. PMBF will be used on the first phase of DLP checkup process to reduce this number to at least thousands or less. The comparison process was done using two approaches to populate PMBF. In first approach one document was placed in one row only but

Fig. 2 Example of source document comparison for the first corpus of 500 documents

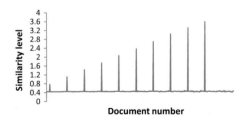

Table 2 False positive and false negative results of the experiment for two approaches to populate PMBF

	Document in one line	Distributed documents
False positive results	301977 ($\sim$1%)	9322 ($\sim$0.01%)
False negative results	51 ($\sim$5%)	117 ($\sim$11%)

each row may contain many documents. In the second approach longer documents were distributed among many lines. As it can be seen from Table 2 second approach leaded to much less false positives but as a trade off for the better compression number of false negatives doubled. Additionally in the second approach each row in the PMBF were populated only up to 50% of its potential capacity to reduce false positives. Both approaches produced some false negative results that are not desirable for DLP systems. Detailed study of false negatives indicated that all of them were caused by the documents that contained less than 1% of the text from the source documents. Moreover 26 documents out of 117 false negatives text inserts from the source documents were highly obfuscated. Since the comparison was done on the word level without any additional tools to tackle text obfuscation, it can be stated that these false positives were expected. Two methods produced different amount of false positives. In the second case when density of BF was reduced by 50% the number of false positives decreased significantly. This feature of PMBF gives DLP developers additional choice—if they are ready to use twice as much memory for PMBF then the second stage of comparison will be much less loaded because of the lower level of false positives.

These two experiments indicated that PMBF can be used in DLP if it is acceptable that only larger portions of the restricted documents will be filtered out. This leaves the attackers with the potential to split the document into the smaller chunks to avoid filtering. This would be very suitable when DLP is intended to protect users from errors such as typos in the email address or other mistakes where users do not have intentions to violate the document distribution rules.

5 Conclusion

The chapter proposed an improved data structure based on Bloom filter to address the issue of fast document comparison. The proposed data structure and algorithms allow better memory allocation in Bloom filters aimed on document comparison. Additional index allows BF to locate the potentially similar documents even if few documents have been placed in a single row of the filter. This index also allows computationally effective document removal operations. The chapter also provided theoretical evaluation of the threshold that could be used to eliminate false positive results of the comparison for two BFs.

One limitation of the proposed approach is related to the fact that using PMBF makes sense only if the entire filter could be allocated in the computer RAM where fast bitwise comparison is possible. Placing parts of the filter on the disk will fade its speed advantage. Based on the experiment above we can state that even minimal server configuration with few Gigabytes of RAM can handle hundreds of thousands of documents which seems to be suitable for DLP systems for a medium enterprise. As first experiment shows PMBF provides noticeable size advantage over matrix BF. Second experiment indicated that PMBFs are applicable for filtering in DLP if document in question includes larger portion (above 5%) of the restricted document. This limitation may not be a problem depending on the purposes of the specific DLP.

Acknowledgements Author would like to acknowledge productive discussions on Bloom Filters with Dr. A. Tskhay, and Mr. V. Shcherbinin as well as help with experiments from Mr. E. Storozhenko, Mr. L. Shi, and Mr. V. Dyagilev.

References

1. Knuth, D., Morris Jr., J., Pratt, V.: Fast pattern matching in strings. SIAM J. Comput. **6**(2), 323–350 (1977)
2. Brin, S., Davis, J., García-Molina, H.: Copy detection mechanisms for digital documents. SIGMOD Rec. **24**(2), 398–409 (1995). doi:10.1145/568271.223855
3. Cormack, G.V.: Email spam filtering: a systematic review. Found. Trends Inf. Retr. **1**(4), 335–455, Apr 2008. doi:10.1561/1500000006 (1995)
4. Butakov, S., Scherbinin, V.: The toolbox for local and global plagiarism detection. Comput. Educ. **52**(4), 781–788, May 2009. doi:10.1016/j.compedu.2008.12.001
5. Bloom, B.H.: Space/time trade-offs in hash coding with allowable errors. Commun. ACM **13**(7), 422–426, July 1970. doi:10.1145/362686.362692
6. Liu, S., Kuhn, R.: Data Loss Prevention. IT Professional, vol. 12, no. 2, pp. 10–13, Mar-Apr 2010. doi:10.1109/MITP.2010.52
7. Blackwell, C.: A security architecture to protect against the insider threat from damage, fraud and theft. In: Proceedings of the 5th Annual Workshop on Cyber Security and Information Intelligence Research (CSIIRW '09), Article 45, p. 4 (2009). doi:10.1145/1558607.1558659
8. Lawton, G.: New technology prevents data leakage. Computer **41**(9), 14–17, Sept 2008. doi:10.1109/MC.2008.394
9. Potter, B.: Document protection: document protection. Netw. Secur. **2008**(9), 13–14, Sept 2008. doi:10.1016/S1353-4858(08)70108-9
10. Wang, J., Xiao, M., Dai, Y.: MBF: a Real Matrix Bloom Filter Representation Method on Dynamic Set, Network and Parallel Computing Workshops, 2007. NPC Workshops, pp. 733–736, 18–21 Sept 2007. doi:10.1109/NPC.2007.107
11. Geravand, S.; Ahmadi, M.A.: Novel adjustable matrix bloom filter-based copy detection system for digital libraries. In: 2011 IEEE 11th International Conference on Computer and Information Technology (CIT), pp. 518–525, 31 Aug 2011–3 Sept 2011. doi:10.1109/CIT.2011.61
12. Guo, D., Wu, J., Chen, H., Luo, X.: Theory and network applications of dynamic bloom filters, INFOCOM 2006. In: Proceedings of the 25th IEEE International Conference on Computer Communications, pp. 1–12, Apr 2006. doi:10.1109/INFOCOM.2006.325
13. Broder, A.Z., Mitzenmacher, M.: Network Applications of Bloom Filters: A Survey. Internet Mathematics, Vol 1, No 4, pp. 485–509 (2005)
14. Fan, L., Cao, P., Almeida, J., Broder, A.: Summary cache: a scalable widearea web cache sharing protocol. IEEE/ACM Trans. Netw. **8**(3), 281–293, (2000)

15. Karp, R.M., Rabin, M.O.: Pattern-matching algorithms. IBM J. Res. Dev. **31**(2), 249–260 (1987)
16. Schleimer, S., Wilkerson, D., Aiken, A.: Winnowing: local algorithms for document finger-printing. In: Proceedings of the ACM SIGMOD International Conference on Management of Data, pp. 76–85, June 2003
17. Potthast, M., Eiselt, A., Stein, B., Barrón-Cedeño, A., Rosso, P.: PAN Plagiarism Corpus PAN-PC-09 (2009). http://www.uni-weimar.de/medien/webis/research/corpora
18. Potthast, M., Gollub, T., Hagen, M., Kiesel, J., Michel, M., Oberländer, A., Stein, B.: Overview of the 4th international competition on plagiarism detection. In: CLEF Online Working Notes/Labs/Workshop (2012)

A Hybrid Approach for the Automatic Extraction of Causal Relations from Text

Antonio Sorgente, Giuseppe Vettigli and Francesco Mele

Abstract This chapter presents an approach for the discovery of causal relations from open domain text in English. The approach is hybrid, indeed it joins rules based and machine learning methodologies in order to combine the advantages of both. The approach first identifies a set of plausible cause-effect pairs through a set of logical rules based on dependencies between words, then it uses Bayesian inference to reduce the number of pairs produced by ambiguous patterns. The SemEval-2010 task 8 dataset challenge has been used to evaluate our model. The results demonstrate the ability of the rules for the relation extraction and the improvements made by the filtering process.

Keywords Natural language processing · Information extraction
Relations extraction · Causal relations

1 Introduction

Causality is an important element for textual inference and the automatic extraction of causal relations plays an important role for the improvement of many Natural Language Processing applications such as question answering [1, 2], document summarization and, in particular, it enables the possibility to reason about the detected events [3, 4]. Also, with the growth of textual unstructured information on web, it is a fundamental tool to provide effective insights about the data. In fact, many websites[1]

---

[1]One of the most prominent examples is http://www.recordedfuture.com/.

A. Sorgente (✉) · G. Vettigli · F. Mele
Institute of Cybernetics "Eduardo Caianiello" of the National Research Council,
Via Campi Flegrei 34, 80078 Pozzuoli (Naples), Italy
e-mail: a.sorgente@cib.na.cnr.it

G. Vettigli
e-mail: g.vettigli@cib.na.cnr.it

F. Mele
e-mail: f.mele@cib.na.cnr.it

C. Lai et al. (eds.), *Emerging Ideas on Information Filtering and Retrieval*,
Studies in Computational Intelligence 746, https://doi.org/10.1007/978-3-319-68392-8_2

specialized in web intelligence provide services for the analysis of huge amounts of texts and in this scenario the extraction of causal information can be used for the creation of new insights and for the support of the predictive analysis.

The automatic extraction of causal relations is also a very difficult task, in particular the English language presents some hard problems for the detection of causal relations. Indeed, there are few explicit lexico-syntactic patterns that are in exact correspondence with a causal relation while there is a huge number of cases that can evoke a causal relation not in a uniquely way. For example, the following sentence contains a causal relation where *from* is the pattern which evokes such relation:

> The most dreaded types of scars are contracture scars **from** burns.

In this case, the words (*scars* and *burns*) connected by the cue pattern (*from*) are in a causal relation while in the following sentence the *from* pattern doesn't evoke the same type of relation:

> *In the past, insulin was extracted **from** the pancreases of cows or pigs.*

Most of the existing approaches [5, 6] to discover causal relations are centered on the extraction of a pair of words or noun phrases that are in a causal relation, they do not discriminate causes and effects.

In this chapter we propose an approach based on a set of rules that uses the dependency relations between the words. It is able to extract the set of potential cause-effect pairs from the sentence, then we use a Bayesian approach to discard the incorrect pairs.

The chapter is organized as follows. In Sect. 2 we present a concise overview of the previous works about causal relations extraction from text. In Sect. 3 we briefly introduce the dependency grammars. Section 4 describes the proposed method. Results are presented in Sect. 5. At the end we offer some discussions and conclusions.

2 Related Works

In this section we will briefly introduce some approaches proposed by other authors concerning the automatic extraction of causal knowledge.

In [6] a method, based on Decision Trees, for the detection of marked and explicit causations has been proposed. The authors showed that their method is able to recognize sentences that contain causal relations with a precision of 98% and a recall of 84%. However, this method is not able to detect the causes and the effects.

With respect to the classification of semantic relations, the SemEval challenge proposes two tasks (task 4 in 2007 [7] and task 8 in 2010 [8]). The tasks concern about the classification of pairs of words, in particular, in each sentence a specific pair of words is already annotated and the target of the tasks consists in classifying the pairs according to the relation evoked in the sentence. The tasks take in account seven types of relations, one of which is the causal relation. In SemEval-2010, Rink and Harabagiu [9] had the best results. They obtained an overall precision of 89%

and an overall recall of 89% using a SVM classifier, for the specific class of the causal relations they obtained a precision of 89% and a recall of 89%.

An approach to identify causes and effects in sentences was proposed in [2] where a semi-automatic method to discover causal relations having the particular pattern $<NPverbNP>$ was defined. The authors reported a precision of 65% on a corpus containing a set of documents related to terrorism.

A system for mining causal relations from Wikipedia is proposed in [5]. In order to select the patterns, the authors used a semi-supervised model based on the dependency relations between the words able to extract pairs of nominals in the causal relations. They reported a precision of 76% and a recall of 85%. Also in this case, the patterns discovered by their algorithm are not able to discriminate the causes from the effects.

In order to predict future events from news, in [10] the authors implemented a method for the extraction of causes and effects. In this case, the domain of interest was restricted to the headlines of newspaper articles and a set of handcrafted rules was used for this task (with a precision of 78%). Finally, in [11] regarding a medical abstracts domain, separated measures of precision and recall for causes and effects are reported: a precision of 46% and a recall of 56% for the causes and a precision of 54% and a recall of 64% for the effects. In the last two works mentioned, the approaches proposed are able to discriminate between causes and effects, but they are limited to particular domains.

3 Some Aspects of Dependency Grammars

Dependency grammars are used to describe the grammatical relations of a sentence [12]. In this grammars, all the words of a sentence, except a specific verb, depend on other words. The verb which doesn't depend on any word is considered the central element of the clause. All the other words are either directly or indirectly dependent on the verb. In this type of segmentation sentence each relation involves two words, one of them is called *head* and the other *dependent*. In this work we have used the Stanford typed dependencies [13], in particular the collapsed version. An example of dependency representation is the following:

$$
\begin{aligned}
&det(fire, The), \\
&nsubjpass(caused, fire), \\
&auxpass(caused, was), \\
&root(ROOT, cause), \\
&det(bombing, the), \\
&agent(cause, bombing).
\end{aligned}
\tag{1}
$$

This is a dependency representation of the sentence "*The fire was caused by the bombing*" (a graphical form of the same representation is showed in Fig. 4 where a dependency link represented as an arrow pointing from the head to the dependent). The dependency structure can be seen as a tree (directed acyclic graph) having the main verb as root. In the example above, *nsubjpass(cause, fire)* indicates that *fire* is the passive subject of the verb *cause*; *auxpass(cause, was)* indicates that *was* is the auxiliary; *root(ROOT, cause)* indicates that *caused* is the root of the dependency tree; *agent(cause, bombing)* indicates that *bombing* is the complement of *caused*; *det(fire, the)* and *det(bombing, the)* represent, respectively, a connection between the words *fire* and *bombing* with their determiners. A complete list of dependency relations has been presented in [13].

4   Extraction Method

In this work, the goal is to extract from a sentence S a set of *cause-effect* pairs $\{(C_1, E_1), (C_2, E_2), \ldots, (C_n, E_n)\}$ where (C_i, E_i) represents the ith cause-effect pair in S. To this end, we propose a method represented in Fig. 1. First, we check if the sentence contains a causal pattern. Then, if a pattern is found the sentence is parsed and a set of rules is applied. Also, a Bayesian classifier is applied to filter out the pairs produced by the rules derived from ambiguous patterns.

4.1   Assumptions

The extraction model has been developed under the following assumptions:

1. The relation is **marked** by a linguistic unit. A sentence contains a marked relation if there is a specific linguist unit that evokes the relation. For example the sentence "*the boy was absent because he was ill*" is marked, indeed there is the unit *because* that evoke the causal relation, while "*Be careful, its unstable*" contains a causal relation but has not a linguist unit.
2. The causation is **explicitly** represented, which means that both arguments of the relation (cause and effect) are present in the sentence. For example the sentence

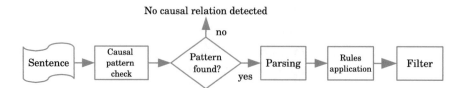

Fig. 1 Steps for the detection of causes and effects of the proposed method

"*I'm a success today because I had a friend who believed in me*" is explicit because contains both cause *I had a friend who believed in me* and the effect *I'm a success*, while the sentence "*John killed Bob*" is implicit because the effect, *the death of Bob*, is not explicitly reported.

3. The arguments of the relation are nominal words denoting:

- an occurrence of an event, a state or an activity;
- a noun as an entity or one of its readings (like *breakfast*, which can denote an entity, or like *devastation*, which can denote an event);
- a metonymy (like *the mall*, which can stand in for shopping).

The assumption 1 and 2 are the basis of the extraction rules, while the last assumption is given by the data available for the analysis.

4.2 Lexico-Syntactic Patterns

In this work, as first step, we have defined a set of lexico-syntactic patterns that represent the structure of the causal relations in the sentence. They represent causal patterns through groups of syntactic structures having common features. In order to identify the lexico-syntactic patterns, we have inspected the structure of the sentences that contain causal relations in the train dataset provided for the task 8 of SemEval-2010.

For this goal, we have analyzed the train dataset studying the χ^2 measure. This statistics measures the dependence between the features and a target variable and it is often used to discover which features are more relevant for statistical classification. The study proposed by Yang and Pedersen in [14] shows that χ^2 is an effective statistics for terms selection. Therefore, we decided to use χ^2 to select the most characterizing patterns associated with the causal relation. We also experimented with the Information Gain statistics and obtained similar results. Analyzing the χ^2 measure we have identified the following groups of patterns:

- *Simple causative verbs*: single verbs having the meaning of "causal action" (e.g. *generate, trigger, make* and so on).
- *Phrasal verbs*: phrases consisting of a verb followed by a particle (e.g. *result in*).
- *Noun + preposition*: expressions composed by a noun followed by a preposition (e.g. *cause of*).
- *Passive causative verbs*: verbs in passive voice followed by the preposition *by* (e.g. *caused by, triggered by*, and so on).
- *Single prepositions*: prepositions that can be used to link "cause" and "effect" (e.g. *from, after*, and so on).

For each lexico-syntactic pattern a regular expression (see Table 1) is defined to recognize the sentences that contain such pattern, and a set of rules is defined in order to detect causes and effects.

Table 1 List of Lexico-syntactic patterns and related regular expression used to detect causal sentence

Pattern	Regular expression
Simple causative verbs	(.*) <cause|generate|triggers|...> (.*)
Phrasal verbs/Noun + preposition	(.*) <result|cause|lead|...> <in|of|to> (.*)
Passive causative verbs	(.*) <caused|generated|triggered|...> by (.*)
Single prepositions	(.*) <from|after|...> (.*)

4.3 Rules

For the extraction of causes and effects in the sentence, the proposed approach uses a set rules based on the relations in the dependency tree of the sentence. For the definition of rules we have analyzed, in the train dataset, the most frequent relations that involve the words labeled as *cause* or *effect* in the dependency tree. This analysis was performed grouping the sentences with respect to the lexico-syntactic patterns. For example, in the case of sentences that match with *phrasal verb* pattern we have observed that the cause is associated to the verb while the effect is associated to the preposition; or in the case of sentences that match with a single preposition both cause and effect are associated to the preposition. The defined rules are formalized by Horn-Clauses. The main rule that allows to detect the cause-effect relation is:

$$cause(S, P, C) \land effect(S, P, E) \rightarrow cRel(S, C, E), \tag{2}$$

where $causeS(S, P, C)$ means that the C is a cause in S in accordance with the pattern P while $effectS(S, P, E)$ means that E is the effect in C with respect to P. So, $cRel(S, C, E)$ represents the causal relation detected in a sentence.

4.3.1 Rules for Simple Causative Verbs

For this pattern, generally the cause and the effect are respectively the subject (rule expressed in Eq. 3) and the object (rule expressed in Eq. 4) of the verb. Examples of verbs which evoke causal relations are *cause, create, make, generate, trigger, produce, emit* and so on. We indicate with $verb(S, P, V)$ that the verb V of the sentence S complies with a pattern P, which has to be a simple causative verb (first row in Table 1), while the relation $nsubj(S, V, C)$ indicates that C is the subject of V in S, and $dobj(S, V, E)$ indicates that E is the direct object of V in S. According to the given definitions, we have formulated the following rules:

$$verb(S, P, V) \land nsubj(S, V, C) \rightarrow cause(S, P, C), \tag{3}$$

$$verb(S, P, V) \land dobj(S, V, E) \rightarrow effect(S, P, E). \tag{4}$$

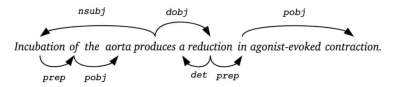

Fig. 2 Dependencies among the words of the sentence "*Incubation of the aorta produces a specific reduction in agonist-evoked contraction*"

If we consider, for example, the dependency tree of the sentence "*Incubation of the aorta produces a specific reduction in agonist-evoked contraction*" (showed in Fig. 2), applying the rules Eqs. 3 and 4, we have that *Incubation* is the cause and *reduction* is the effect.

4.3.2 Rules for Phrasal Verbs/Noun + Preposition

For this pattern, the cause is linked to the verb (or noun) while the effect is linked to the preposition. We indicate with $prep\_verb(S, P, V)$ that the verb or the noun V of the sentence S complies to the pattern P, which has to be a phrasal verb (second row in Table 1). While, $prep(S, E, Pr)$ indicates that Pr is a propositional modifier of V in S. According to the given definitions, we have formulated the following rules:

$$prep\_verb(S, P, V) \wedge nsubj(S, V, C) \rightarrow cause(S, P, C), \qquad (5)$$

$$prep\_verb(S, P, V) \wedge preposition\_of(S, V, Pr) \wedge$$
$$\wedge\ prep(S, E, Pr) \rightarrow effect(S, P, E). \qquad (6)$$

In the particular case where the causal relation is introduced by the expression "**due to**" we have that, with respect to other expressions that are complied to this pattern, the terms of the relations are *inverted*. So, the cause is linked to the preposition and the effect to the verb. The rules are:

$$prep\_verb(S, P, due) \wedge prep(S, due, to) \wedge pobj(S, to, C) \rightarrow cause(S, P, C),$$
$$prep\_verb(S, P, due) \wedge nsubj(S, due, E) \rightarrow effect(S, P, E),$$
$$\qquad (7)$$

where $pobj(S, Pr, C)$ indicates that C is the object of the preposition Pr in S. In this case, applying the rules Eq. 7 on the dependency tree of the sentence "*A secret for avoiding weight gain due to stress is the use of adaptogens*" (showed in Fig. 3), we are able to correctly detect *stress* as cause and *gain* as effect.

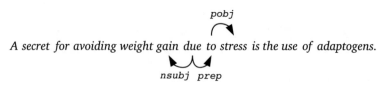

A secret for avoiding weight gain due to stress is the use of adaptogens.

Fig. 3 Dependencies among the words of the sentence "*A secret for avoiding weight gain due to stress is the use of adaptogens*"

4.3.3   Rules for Passive Causative Verbs

In this pattern, the cause is the word of the sentence that has an *agent* relation with the verb. In fact, as reported in [15], the *agent* is the complement of a passive verb which is introduced by the preposition *by* and does the action, while the effect is the passive subject of the verb. We indicate with $passive(S, P, V)$ that the verb V of the sentence S complies with the pattern P, which has to be a passive causative verb (third row in Table 1). According to the given definitions, we have formulated the following rules:

$$passive(S, P, V) \wedge agent(S, V, C) \rightarrow cause(S, P, C),$$
$$passive(S, P, V) \wedge nsubjpass(S, V, E) \rightarrow effect(S, P, E), \tag{8}$$

where $agent(S, V, C)$ indicates that C is the complement of a passive verb V and $nsubjpass(S, V, E)$ indicates that E is the subject of V in S.

In this case, applying the rules on the dependency tree of the sentence "*The fire was caused by the bombing.*" (showed in Fig. 4), we are able to correctly detect *bombing* as cause and *fire* as effect.

4.3.4   Rules for Single Prepositions

For this pattern, the cause and effect are linked, in the dependence tree, to the preposition that evokes the causal relation.

We use $preposition(S, P, Pr)$ to indicate that the preposition Pr of the sentence S complies with the pattern P, which has to be a single preposition (forth row in Table 1). In this case, we have formulated the following rules:

Fig. 4 Dependencies among the words of the sentence "*The fire was caused by the bombing*"

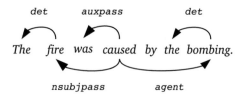

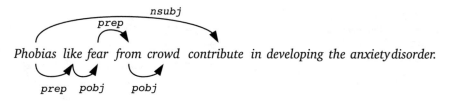

Fig. 5 Dependencies among the words of the sentence "*Phobias like fear from crowd contribute in developing the anxiety disorder*"

$$preposition(S, P, Pr) \wedge pobj(S, Pr, C) \rightarrow cause(S, P, C),$$
$$preposition(S, P, Pr) \wedge prep(S, E, Pr) \rightarrow effect(S, P, E).$$
(9)

In many cases the effects have a direct link with the verb that precedes *from* or *after*. In order to handle those situations we defined the following rule:

$$preposition(S, P, Pr) \wedge prep(S, V, Pr) \wedge nsubj(S, V, C) \rightarrow effect(S, P, C).$$
(10)

If we consider, for example, the dependency tree of the sentence "*Phobias like fear from crowd contribute in developing the anxiety disorder*" (showed in Fig. 5), applying the rules Eqs. 9 and 10, we have that *crow* is the cause and *fear* is the effect.

4.3.5 Rules for Multiple Causes and Effects

The rules presented above allow to detect a cause and an effect for each pattern that matches in a sentence. If there are two or more causes for an effect, we want to detect them all. For example, in the sentence

Heat, wind and smoke cause flight delays.

the *and* relation indicates that the *delays* (effect) is caused by *Heat*, *wind* and *smoke*, so we have three causes. To deal with these situations we have defined rules that propagate the causal relation along the conjunct dependencies:

$$cause(S, P, C1) \wedge conj(S, C1, C2) \rightarrow cause(S, P, C2).$$
(11)

A similar rule is defined to propagate through conjunctions the effect:

$$effect(S, P, E1) \wedge conj(S, E1, E2) \rightarrow effect(S, P, E2),$$
(12)

where $conj(S, C1, C2)$ indicates that $C1$ and $C2$ are connected by a coordinating conjunction (*and, or*).

The Fig. 6 shows the dependency tree of sentence "*Heat, wind and smoke cause flight delays*" and we can see that applying these rules we detect all the causes.

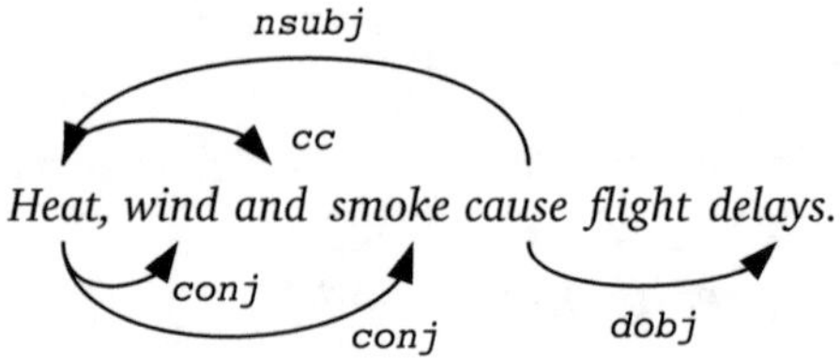

Fig. 6 Dependencies among the words of the sentence "*Heat, wind and smoke cause flight delays*"

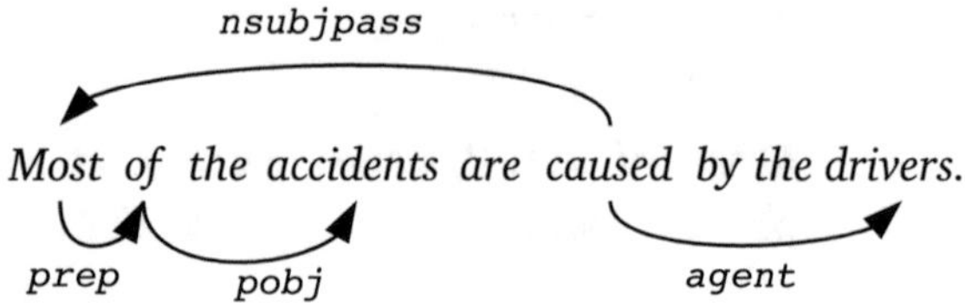

Fig. 7 Dependencies among the words of the sentence "*Most of the accidents are caused by the drivers*"

4.3.6 Rules for Partitives

During our test, we have observed that the rules presented below extract erroneous causes or effects when they are involved in the partitive of a sentence. The partitive denote a grammatical construction used to indicate that only a part of a whole is referred to. It usually has the form *[DP Det. + of + [DP Det. + NP]]*, the first determiner is a quantifier word that quantifies over a subset or part of the embedded determiner phrase (DP), which either denotes a set or a whole respectively. The second determiner is usually an article, a demonstrative, a possessive determiner, or a quantifier. For example *most of the students, some of the children, a slice of cake* and so on. In this cases the rules detect as cause or effect the quantifier, for example if we take in account the sentence showed in Fig. 7, applying the rules in Eq. 8, we extract as effect the word *Most*, while the correct word *accidents*. To overcame this problem, we have defined the following rules:

$$
\begin{aligned}
cause(S, P, C) &\wedge quantifier(C) \wedge \\
&prep(S, C, of) \wedge pobj(S, of, X) \rightarrow cause(S, P, X), \\
effect(S, P, E) &\wedge quantifier(E) \wedge \\
&prep(S, E, of) \wedge pobj(S, of, X) \rightarrow effect(S, P, X).
\end{aligned}
\tag{13}
$$

These rules are applied after the other rules introduced above. So, in the case where the cause (or the effect) identified is a quantifier, the partitive structure mentioned above is inspected to identify the object of the preposition *of*. Indeed, applying the rules in Eq. 13 after the ones in Eq. 8, we are able to correctly extract the effect from the sentence in Fig. 7.

4.4 Pairs Filtering

Due to their empirical nature, the patterns and the rules defined above are not able to produce exact results. Hence, some pairs that are not in causal relation can be extracted. In order to remove the erroneous pairs detected we used a binary classifier to discriminate *causal* and *non-causal* pairs. This problem is a subtask of the task 8 of the SemEval-2010 where only the causal relation has been considered. To implement the classifier we have chosen to use the Bayesian classification method. Considering the (hypothetical causal) pair $r \equiv cRel(S, C, E)$, the Bayes' rule becomes:

$$P(c_i|r) = \frac{P(r|c_i)P(c_i)}{P(r)}, \qquad (14)$$

with $i = 1, 2$ where c_1 is *causal* and c_2 is *non-causal*. The following features have been associated to the relation r:

- *Lexical features*. The words between C and E.
- *Semantic features*. All the hyponyms and the synonyms of each sense of C and E reported in WordNet [16].
- *Dependency features*. The direct dependencies of C and E in the dependency parse tree.

For each pair r we have extracted a set of features F and for each feature $f \in F$ we have estimated $P(f|c_i)$ by counting the number of causal relations having the feature f, then dividing by the total number of times that f appears. We have used Laplace smoothing applying an additive constant α to allow the assignment of non-zero probabilities for features which do not occur in the train data:

$$P(f|c_i) = \frac{\#(c_i, f) + \alpha}{\#f + \alpha|F|}. \qquad (15)$$

Assuming that the features are independent from each other we computed

$$P(r|c_i) = \prod_{f \in F} P(f|c_i). \qquad (16)$$

According to the Bayesian classification rule, the relation is classified as *causal* if

$$P(c_1|r) \geq P(c_2|r) \qquad (17)$$

and as *non-causal* otherwise.

The classifier discussed in this section has been tested with 10-folds cross validation on the train set of the SemEval-2010 dataset in order to select a value for the parameter α. Then, the classifier has been tested on the testset. We were able to achieve a precision of 91% and a recall of 88%, which slightly improve the results achieved during the SemEval challenge (87% precision, 87% recall).

5 Evaluation

We have evaluated our method on a test corpus made extending the annotations of the SemEval-2010 (Task 8) test set. In the original dataset in each sentence only one causal pair has been annotated. We have extended the annotation with the causal pairs not considered by the SemEval annotators. In the cases where an effect is caused by a combination of events or a cause produces a combination of events, pair cause-effect is annotated separately. Our corpus is composed by 600 sentences, 300 of them contain at least a causal relation and the other 300 without causal relations.

The dependency trees have been computed using the Stanford Statistical Parser [17] and the rules for the detection of cause-effect pairs have been implemented in XSB Prolog [18].

The performances have been measured globally and per sentence. The metrics used are *precision*, *recall* and *F-score* in both contexts. Let us define *precision* and *recall* in the global context as

$$P_{global} = \frac{\#\text{correct retrieved pairs}}{\#\text{retrieved pairs}}, \tag{18}$$

$$R_{global} = \frac{\#\text{correct retrieved pairs}}{\#\text{total pairs in } D}, \tag{19}$$

where D is the set of all the sentences in the dataset. The *precision* and the *recall* to measure the performances *per sentence* are defined as

$$P_{sentence} = \frac{1}{|M|} \sum_{s \in M} \frac{\#\text{correct retrieved pairs in } s}{\#\text{retrieved pairs in } s}, \tag{20}$$

$$R_{sentence} = \frac{1}{|D|} \sum_{s \in D} \frac{\#\text{correct retrieved pairs in } s}{\#\text{total pairs in } s}, \tag{21}$$

where M is the set of the sentences where the rules found at least a causal pair. In both cases the F-score is defined as

$$F = 2\frac{P \cdot R}{P + R}. \tag{22}$$

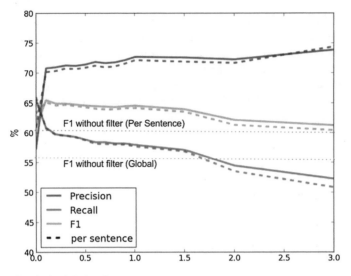

Fig. 8 Results obtained during the tests

The *per sentence* metrics measure the ability of the system to extract all the causal pairs contained in a given sentence while the *global* metrics measure the ability of the system to extract all the causal pairs contained in the entire corpus.

In order to evaluate the proposed model we computed the metrics described above, first, using only the rules, and then, combining the rules and the pair filter with different values of the parameter α.

The results of the evaluation are summarized in Fig. 8. We can see that the global F-score of the rules stand-alone is around 55% and the per sentence F-score is around 60% (see dotted lines Fig. 8). While, we can see that the application of the filter always increases the F-score. The best performances of the system are obtained with Laplace smoothing setting $\alpha = 0.2$. For higher values of α we are able to achieve a higher precision, but we also have a significant lowering of the recall.

6 Conclusion and Future Work

We have presented an approach for the detection and the extraction of cause-effect pairs in English sentences. With respect to the other works presented in Sect. 2, the task addressed in this chapter is more detailed because our approach focuses on the extraction of all the cause-effect pairs in a causal sentences.

The proposed method combines rules, for the extraction of all possible causes and effects, and a Bayesian classifier to filter the erroneous solutions. Such approach works on sentences that contain explicit and marked causal relation.

The evaluation of the proposed method has been done on an extended version of the testset used for the task 8 of the SemEval-2010 challenge. The results achieved by our approach are encouraging, especially if we consider that the dataset contains data extracted from various domains.

Also, we believe that the rules applied by the model can be easily adapted for other information extraction tasks and that they provide interesting insights on how explicit causal relation are expressed under the point of view of the grammatical dependencies.

In the future, we plan to refine the rules presented and experiment other filtering techniques. In order to evaluate the approach we also plan to extend the dataset developed in this work. Also, we will extend this system in order to handle implicit causal relations and casual relation don't marked by an explicit lexical unit.

References

1. Atzeni, P., Basili, R., Hansen, D.H., Missier, P., Paggio, P., Pazienza, M.T., Zanzotto, F.M.: Ontology-based question answering in a federation of university sites: the moses case study. In: NLDB. 413–420 (2004)
2. Girju, R., Moldovan, D.: Mining answers for causation questions. In: AAAI symposium on (2002)
3. Mele, F., Sorgente, A.: OntoTimeFL—a formalism for temporal annotation and reasoning for natural language text. In: Lai, C., Semeraro, G., Vargiu, E. (eds.) New Challenges in Distributed Information Filtering and Retrieval. Studies in Computational Intelligence, vol. 439, pp. 151–170. Springer, Berlin (2013)
4. Mele, F., Sorgente, A., Vettigli, G.: Designing and building multimedia cultural stories using concepts of film theories and logic programming. In: AAAI Fall Symposium: Cognitive and Metacognitive Educational Systems (2010)
5. Ittoo, A., Bouma, G.: Extracting explicit and implicit causal relations from sparse, domain-specific texts. In: Proceedings of the 16th International Conference on Natural Language Processing and Information Systems. NLDB'11, pp. 52–63. Springer, Berlin (2011)
6. Blanco, E., Castell, N., Moldovan, D.I.: Causal relation extraction. In: LREC (2008)
7. Girju, R., Nakov, P., Nastase, V., Szpakowicz, S., Turney, P., Yuret, D.: Classification of semantic relations between nominals. Lang. Resour. Eval. **43**(2), 105–121 (2009)
8. Hendrickx, I., Kim, S.N., Kozareva, Z., Nakov, P., Saghdha, D., Romano, L., Szpakowicz, S.: Semeval-2010 task 8: multi-way classification of semantic relations between pairs of nominals
9. Rink, B., Harabagiu, S.: Utd: Classifying semantic relations by combining lexical and semantic resources. In: Proceedings of the 5th International Workshop on Semantic Evaluation. SemEval '10, Stroudsburg, PA, USA, Association for Computational Linguistics (2010) 256–259
10. Radinsky, K., Davidovich, S.: Learning to predict from textual data. J. Artif. Int. Res. **45**(1), 641–684 (2012)
11. Khoo, C.S.G., Chan, S., Niu, Y.: Extracting causal knowledge from a medical database using graphical patterns. In: Proceedings of the 38th Annual Meeting on Association for Computational Linguistics. ACL '00, Stroudsburg, PA, USA, pp. 336–343. Association for Computational Linguistics (2000)
12. Kruijff, G.J.: Dependency grammar. In: Brown, K. (ed.) Encyclopedia of Language and Linguistics (Second Edition), 2nd edn, pp. 444–450. Elsevier, Oxford (2006)
13. Covington, M.A.: A fundamental algorithm for dependency parsing. In: Proceedings of the 39th Annual ACM Southeast Conference, pp. 95–102 (2001)

14. Yang, Y., Pedersen, J.O.: A comparative study on feature selection in text categorization. In: Proceedings of the Fourteenth International Conference on Machine Learning. ICML '97, San Francisco, CA, USA, pp. 412–420. Morgan Kaufmann Publishers Inc. (1997)
15. Marneffe, M.C.D., Manning, C.D.: Stanford Typed Dependencies Manual (2008)
16. Fellbaum, C.: WordNet: An Electronic Lexical Database. Bradford Books (1998)
17. De Marneffe, M.C., Maccartney, B., Manning, C.D.: Generating typed dependency parses from phrase structure parses. In: LREC 2006 (2006)
18. Sagonas, K., Swift, T., Warren, D.S.: Xsb as an efficient deductive database engine. In: Proceedings of the ACM SIGMOD International Conference on the Management of Data, pp. 442–453. ACM Press (1994)

Ambient Intelligence by ATML Rules in BackHome

Eloi Casals, José Alejandro Cordero, Stefan Dauwalder, Juan Manuel Fernández, Marc Solà, Eloisa Vargiu and Felip Miralles

Abstract Applying ambient intelligence to assistive technologies and remote monitoring systems is a challenging task, especially in the field of healthcare. In this chapter, we focus on how ambient intelligence has been implemented within the telemonitoring and home support system developed in BackHome, an EU project that aims to study the transition from the hospital to the home for people with disabilities. The proposed solution adopts a rule-based approach to handle efficiently the system adaptation, personalization, alarm triggering, and control over the environment. Rules are written in a suitable formal language, defined within the project, and can be automatically generated by the system taking into account dynamic context conditions and user's habits and preferences. Moreover, they can be manually defined by therapists and caregivers to address specific needs. In this chapter, we also present the concept of "virtual device", which is a software element that mashes together information from two or more sensors in order to make some inference and provide new information. The abstraction of sensor information creates a powerful

Eloi Casals, José Alejandro Cordero, Stefan Dauwalder, Juan Manuel Fernández, Marc Solà, Eloisa Vargiu contributed equally to this work.

E. Casals · J.A. Cordero · M. Solà
Barcelona Digital Technology Center, Barcelona, Spain
e-mail: ecasals@bdigital.org

J.A. Cordero
e-mail: jacordero@bdigital.org

M. Solà
e-mail: msola@bdigital.org

S. Dauwalder · J.M. Fernández · E. Vargiu (✉) · F. Miralles
Eurecat Technology Center, eHealth Unit, Barcelona, Spain
e-mail: eloisa.vargiu@eurecat.org

S. Dauwalder
e-mail: stefan.dauwalder@eurecat.org

J.M. Fernández
e-mail: juanmanuel.fernandez@eurecat.org

F. Miralles
e-mail: felip.miralles@eurecat.org

© Springer International Publishing AG 2018
C. Lai et al. (eds.), *Emerging Ideas on Information Filtering and Retrieval*,
Studies in Computational Intelligence 746, https://doi.org/10.1007/978-3-319-68392-8_3

layered system, which provides a simpler, more readable and more natural description of the rules. Thanks to the proposed solution, end-users can rely on the automatic customization of their user interfaces according to their context and preferences; caregivers can personalize the system to the user's need; and therapists can monitor user's activities easily and can be informed as soon as an alarm is triggered or an anomaly detected.

Keywords Ambient intelligence · Markup languages · Assistive living Telemonitoring

1 Introduction

Ambient Intelligence (AmI) is a term coined by Philips management to conjure up a vision of an imminent future in which persons are surrounded by a multitude of fine-grained distributed networks comprising sensors, computational devices and electronics that are unobtrusively embedded in everyday objects such as furniture, clothes, and vehicles, and that together create electronic habitats that are sensitive, adaptive and responsive to the presence of people [1].

Ambient in AmI refers to the environment and reflects the need for an embedding of technology in a way that it becomes unobtrusively integrated into everyday objects. *Intelligence* reflects that the digital surroundings exhibit specific forms of social interaction, i.e., the environments should be able to recognize the people that live in it, adapt themselves to them, learn from their behaviour, and possibly act upon their behalf [2].

In this chapter, we focus on the application of AmI on assistive technology and remote monitoring. In fact, in this chapter we present how AmI is currently provided and implemented in BackHome. BackHome is an EU project concerning physical and social autonomy of people with disabilities, by using mainly Brain/Neural Computer Interface (BNCI) and integrating other assistive technologies as well [17].

AmI is essential within BackHome since people with severe disabilities could benefit very much from the inclusion of pervasive and context-aware technologies. AmI is, due to its assistive nature, very beneficial for assistive applications. Among the most common types of assistance, the project takes into account the enhancement in the capabilities of controlling some environment features and appliances, as well as the understanding of situations to proactively perform actions. Thus, the physical environment has been enhanced with smart-nodes, smart objects, as well as user's wearable sensors. Adaptation, personalization, alarm triggering, and control over environment are handled with a rule-based approach that relies on a suitable language, defined within the project.

The chapter is organized as follows, Sect. 2 resumes main related work concerning ambient intelligence, its application to assistive techonology and telemonitoring and how trigger actions work in ambient intelligence systems. In Sect. 3, we summarize the objective and the aim of the BackHome project, in which ambient intelligence features are provided to support people with severe disabilities. Section 4 presents the role of ambient intelligence in BackHome. In Sect. 5, the current implementation of ambient intelligence in BackHome is presented; whereas in Sect. 6 a real scenario is illustrated to better clarify all the approach. Section 7 ends the chapter with some conclusions and future work.

2 Ambient Intelligence

According to Augusto et al. [5] we may see Ambient Intelligence (AmI) as the confluence of Pervasive Computing (aka, Ubiquitous Computing), Artificial Intelligence (AI), Human Computer Interaction (HCI), Sensors, and Networks. First, an AmI system pervasively senses the environment by relying on a network of sensors. The gathered information is, then, processed by AI techniques to provide suitable actions to be performed on the environment through controllers and/or specialized HCI.

According to [14], the AmI is placed in the confluence of a multi-disciplinary and heterogeneous ecosystem. This position allows the AmI applications to get the information of the surroundings, actuate and change the environment, the different human interfaces available, as well as apply some reasoning techniques. The conjunction of all these fields is used by the AmI systems always respecting the privacy of the user. Following this description, a system incorporates AmI principles if the following characteristics are met: *Sensitive*, AmI systems have to incorporate the ability to perceive their immediate surroundings; *Responsive*, AmI systems have to be able to react in front of the context occurring; *Adaptive*, AmI systems are to be flexible enough to accommodate the responses along the time; *Transparent*, AmI systems have to be designed to be unobtrusive; *Ubiquitous*, AmI systems have to be concealed so as to minimize the impact of bulky and tedious appliances; and *Intelligent*, AmI systems have to incorporate intelligent algorithms to react in front of specific scenarios.

2.1 Application to Assistive Technology and Telemonitoring

Being interested in using AmI solutions to assist and remote monitor people, among the huge number of application fields in which AmI is currently studied and applied, we focus on assistive technology and telemonitoring.

The capability of AmI techniques for recognizing activities [6, 25], monitoring diet and exercise [19], and detecting changes or anomalies [15] support the key idea of providing help to individuals with cognitive or physical impairments. For instance, AmI techniques can be used to provide reminders of normal tasks or the step sequences to properly realize and complete these tasks. For those with physical limitations, automation and inclusion of AI to their home and work environment may become a response for independent living at home [36].

There have been several other research projects that have investigated the use of various sensor technologies on monitoring of daily activity [12, 13, 27, 28, 35]. More recently, Barger et al. [7] examine whether a system of basic motion sensors can detect behavioral patterns. They adopt a mixture model framework to develop a probabilistic model of behavior. The effect of analyzing behavior during work and off days separately is also examined. Zhou et al. [38] propose a case-driven AmI (C-AmI) system, aiming at sensing, predicting, reasoning, and acting in response to the elderly activities of daily living at home. The C-AmI system architecture is developed by synthesizing various sensors, activity recognition, case-based reasoning, along with Elderly in-Home Assistance (EHA) customized knowledge, within a coherent framework. ALZ-MAS (ALZheimer Multi-Agent System) [32] is a distributed multi-agent system designed upon AmI and aimed at enhancing the assistance and health care for Alzheimer patients living in geriatric residences. Agents in ALZ-MAS collaborate with context-aware agents that employ Radio Frequency Identification (RFID), wireless networks, and automation devices to provide automatic and real time information about the environment, and allow the users to interact with their surroundings, controlling and managing physical services. In [31], authors describe a system based on a wireless sensor network, in order to detect events in a context of a smart home application. The system is addressed to assist in an automatic manner elderly people detecting anomalous event and alerting medical caretakers. The posture of the people is monitored with the help of different cameras and an accelerometer worn by users.

As for healthcare telemonitoring, Corchado et al. [16] describe a telemonitoring system aimed at improving healthcare and assistance to dependent people at their homes. The system proposed in [10] has been designed to provide users personalized health care services through AmI. The system is responsible of collecting the information about the environment considered important, the study of the behavior of the user in terms of the management of the house and inferring of rules for that management. The information collected may be the temperature, luminosity or humidity of the rooms, the position of the user or other persons inside the house, and the state of the required home appliances. An enhancement of the monitoring capabilities is achieved by adding portable measurement devices worn by the user thus vital data is also collected out of the house. Some reactive and simple decisions can be taken by the system when some predetermined events occur.

2.2   Rule-Based Solutions in Ambient Intelligence

Being implemented in real-world environments, AmI involves problems such as incompleteness and uncertainty of the information available about the user and the environment. In fact, we generally deal with information that might be in some way correct, or partly incorrect or missing. Thus, an elaborated reasoning process that deals with those information drawbacks must be performed to successfully define an accurate knowledge representation. To this end, AmI relies on the context as a model of the current situation of the user and its immediate environment [18].

Current implementations of AmI in the field of Smart Home and assistive technologies are focused on providing modifiable assistance according to the user context, the so-called personalized assistance. The more frequent implementations are based on machine learning algorithms and intelligence systems. Nevertheless, this is not the only possible approach: there exist systems based on the use of rule based engines that determine the actions to be triggered by the system. Acampora and Loia [3] present a distributed AmI system, based on agents, communication protocols (TCP/IP) and Fuzzy Logic [37] using as a language for description of knowledge and rules, the Fuzzy Markup Language [4]. The platform DOAPAmI [22] uses a Domain Specific Language (DSL) in order to define the complete platform including services, sensors, profiles of physical platforms where executing the system. This DSL also allows the specification of rules for the different situations, but focused on the proper running of the system not on the user needs and preferences. Papamarkos et al. [29] present an Even-Condition-Action centred approach based on RuleML [23] and focused on Semantic Web, far from the focus of our objective. Other example of use of XML languages in order to define rules based on the context was presented by Schmidt [30] introducing an extension of the Standard Generalized Markup Language (SGML) defined in [8, 9]. This extension allows the language to define triggers and represent the context to improve the user interface in small devices.

3   The BackHome Project in a Nutshell

BackHome is an EU project concerning physical and social autonomy of people with disabilities, by using mainly BNCI and integrating other assistive technologies as well.

BackHome is partly based on the outcomes coming from BrainAble, san EU project aimed at offering an ICT-based human-computer-interaction composed of BNCI system combined with affective computing, virtual environments and the possibility to control heterogeneous devices like smart home environments and social networks [26]. BackHome advances BrainAble in supporting the transition from institutional care to home post rehabilitation and discharge [17].

BackHome aims to study the transition from the hospital to the home, focusing on how people use BNCIs in both settings. Moreover, it is aimed to learn how different BNCIs and other assistive technologies work together and can help clinicians, disabled people, and their families in the transition from the hospital to the home. The final goal of BackHome is to reduce the cost and hassle of the transition from the hospital to the home by developing improved products. To produce applied results, BackHome will provide: new and better integrated practical electrodes; friendlier and more flexible BNCI software; and better telemonitoring and home support tools.

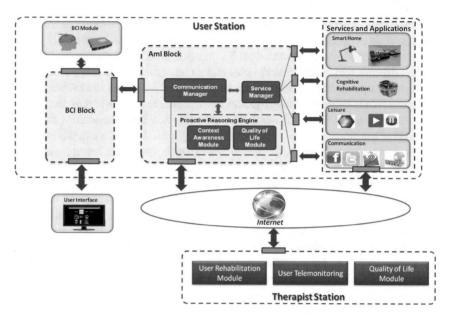

Fig. 1 The architecture of the BackHome platform

Among the overall provided functionalities, in this chapter, we are mainly focusing on how to provide telemonitoring by a sensor-based system and home support by smart home devices.

The BackHome platform, depicted in Fig. 1, relies on two stations: (i) the therapist station and (ii) the user station. The therapist station is focused on offering information and services to the therapists via a usable and intuitive user interface. It is a Web application that allows the therapist to access the information of the user independently of the platform and the device. This flexibility is important in order to get the maximum potential out of the telemonitoring because the therapist can be informed at any moment with any device that is connected to the Internet (PC, a smart phone or a tablet). The user station is the main component that the user interacts with. It contains the modules responsible for the user interface, the intelligence of the system, as well as to provide all the services and functionalities of BackHome. The user station will be completely integrated into the home of the user together with the assistive technology to enable execution and control of these functionalities.

4   Ambient Intelligence in BackHome

AmI is essential within BackHome since people with functional diversity could benefit very much from the inclusion of pervasive and context-aware technologies.

In fact, as already discussed, AmI is, due to its assistive nature, very beneficial for assistive applications.

Among the most common types of assistance, we take into account the enhancement in the capabilities of controlling some environment features and appliances, as well as the understanding of situations to proactively perform actions such as triggering an alarm in case of emergency. In particular, the physical environment will be enhanced by integrating information on user's physical and social environment; information of objects augmented with sensing computing and networking capabilities; as well as wearable sensors. In fact, users wear a BCI system that allows monitoring EEG, EOG, and EMG. Further physiological sensors could be considered, such as ECG and hearth rate, respiration rate, GSR, EMG switches, and inertial sensors. Those sensors allow to monitor mood, health-status, mobility, usual activities, as well as pain and discomfort [33].

Adaptation, personalization, alarm triggering, and control over the environment are handled with a rule-based approach. To this end, a suitable language has been defined. Rules are automatically generated by the systems according to the given context and depending on user's habits and preferences. Moreover, rules can be manually defined by therapists and caregivers.

4.1 The Sensor-Based System

To monitor users at home, we developed a sensor-based telemonitoring and home support system (TMHSS) able to monitor the evolution of the user's daily life activity [34].

The implemented TMHSS is able to monitor indoor activities by relying on a set of home automation sensors. More precisely, we use motion sensors, to identify the room where the user is located (one sensor for each monitored room); a door sensor, to detect when the user enters or exits the premises; electrical power meters and switches, to control leisure activities (e.g., television and pc); pressure sensors, to track user transitions between rooms; and bed (seat) sensors, to measure the time spent in bed (wheelchair). Figure 2 shows an example of a home with the proposed sensor-based system.

From a technological point of view, we use wireless z-wave sensors that send the retrieved data to a central unit located at user's home. That central unit collects all the retrieved data and sends them to the cloud where they will be processed, mined, and analyzed.

Besides real sensors, the system also comprises "virtual devices". Virtual devices are software elements that mash together information from two or more sensors in order to make some inference and provide new information. In so doing, the TMHSS is able to perform more actions and to be more adaptable to the context and the user's habits. Furthermore, the mesh of information coming from different sensors can provide useful information to the therapist. In other words, the aim of a virtual device is to provide useful information to track the activities and habits of

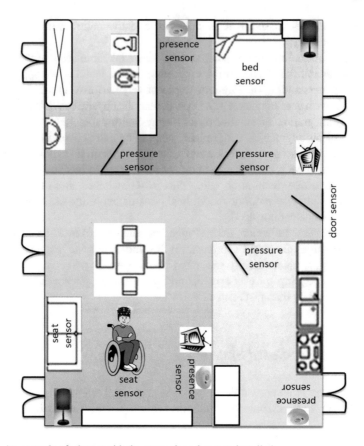

Fig. 2 An example of a home with the sensor-based system installed

the user, to send them back to the therapist through the therapist station, and to adapt the user station, with particular reference to its user interface, accordingly.

4.2 Rule Definition and Triggering

In order to establish a platform-independent and flexible definition of rules, and to specify conditions and actions to be triggered (hereinafter, triggers), in BackHome we adopt a human readable XML-based language called AmI Triggering Markup Language (ATML) [21]. This approach allows exporting the rules definitions to any AmI system and reaching the same status on the configuration of the intelligence.

Rules can be automatically defined by the system depending on the context. Moreover, rules can be manually defined by caregivers and therapists through suitable user interfaces.

4.2.1 The ATML

For the sake of completeness, let us summarize here the adopted language. ATML is compliant with the RuleML in the sense that most of ATML definitions rely on the RuleML definitions. A trigger in ATML is described by its *name*—in order to distinguish it from the overall set of triggers—and a set of *properties*—each one incorporating different attributes helping to understand how to interact with the rule. Moreover, a rule defines the implications of its action through two main sub-sections: *head* and *body*. *Head* is used to define the actions (*TriggerAction*) to be executed whenever the condition of the rule is met. Every *TriggerAction* is defined by a single action to be performed when a given condition (defined in the body section) is met. A *TriggerAction* is defined through three tags: *op*, which indicates the command to be performed on the target; *who*, which describes which system performs the operation; and *device*, which defines the targeted appliance. *Body* expresses the condition of the rule by means of comparison and logic operation.

4.2.2 Context-Aware Rule Definition

BackHome takes advantage of AmI to provide personalization and adaptation. In particular, starting from the information of the context, triggers and rules may be defined. A knowledge representation of the context which needs to be captured, and stored, from the different data/information sources and adopted devices, has been devised. The outcome of the context formalization is depicted in Fig. 3, in which the different context component are presented. This definition incorporates different categories taken into account when evaluating the context:

- Time: representing the current moment taking place.
- Environment: referring to direct information of the context depending on environmental measures.
- Habits: providing information about physiological measurements and normal activities.
- Device: representing the status of the devices controlled by the system.

To perform personalization and adaptation we rely on machine learning techniques able to infer the behavioral patterns of the system, to learn user's habits and to adapt according to user's preferences. Currently, the proposals of the BrainAble [20] project based on AdaBoost with C45 as a weak algorithm [11] has been adopted.

5 The Current Prototype

BackHome is an ongoing project, currently at the end of its second year. BackHome end-users are located in Würzburg and Belfast.

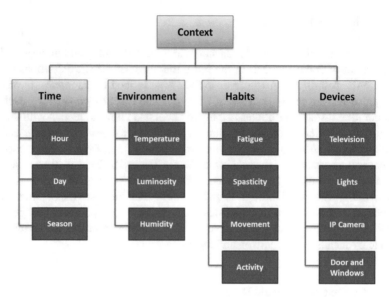

Fig. 3 Context definition in BackHome

Currently, a prototype of the user station has been installed and is currently under testing in both end-user locations. The user station can be controlled by the BCI through a novel speller and provides the following services: smart home control, Web browsing, e-mail managing (GMail), Facebook, Twitter, a multimedia player, cognitive rehabilitation tasks, and the Brain Painting [24].

The TMHSS is not yet integrated with the overall system and is currently under testing in two healthy-users' homes, one in Barcelona and one in Belfast (see Sect. 5.1).

The current version of the therapist station is online and used by therapists, care-givers, and researchers at both end-user locations.

As already said, rules can be automatically defined according to the context (see Sect. 4.2.2). Moreover, they can be defined by caregivers and therapists through suitable interfaces, which generate automatically the corresponding ATML code.

5.1 The Current Sensor-Based System

Several sensors have been currently installed at the selected healthy-user's homes; the configuration of the sensor-net depending on the real map of the homes.

Basic sensors are: motion sensors and electric power devices. The former are triggered when a movement is detected. A motion sensor is placed in each room to detect the movement of the user and the room where s/he is staying. Let us note that, as with alarm systems, it is important to find the optimal placement of these sensors, in order to detect correctly movements and presence, and avoid false detection. The latter are plugged in between the power supply socket and the power plug. Several home appliances and devices can be then controlled remotely by relying on those sensors; such as, televisions, PCs, game consoles, kettles, microwaves.

Apart from those basic sensors, additional sensors have been taken into account. In the current implementation, in fact, we are using mat sensors, to detect when the user is lying in bed or sitting on the sofa. Moreover, we have installed door and window sensors that are magnetic devices able to detect door and/or window openings. Finally, environmental sensors, such as temperature, humidity and luminosity, have been installed to get information about the environment and, thus, the context in which the user is.

Besides the real sensors listed above, two virtual devices have also been developed. The function of the former is to provide information about the position of the user at home. That virtual device mashes information coming from all the motion sensors, as well as the mat ones, to infer the position of the user in every moment. In so doing, the TMHSS is able to determine the room where the user spends the majority of his time, as well as how many minutes/hours s/he stays in each room. Thus, together with the activities registered by the electrical power devices, the TMHSS can make inferences regarding the context, in terms of environment and user's habits. The function of the latter is to aggregate the control of several actuators using an abstraction. In so doing, the TMHSS is able to access to the composing actuators as a single entity.

Let us note that just by using a virtual device, the manual definition of rules is already simplified. To clarify that point, consider the following rule that the caregiver would have to write to turn off main appliances (i.e., TV and lights) if the user is not at home.

```
<Trigger>
   <name>Electrical devices OFF</name>
   <Implies>
      <head>
         <TriggerAction>
            <op>
               <rel>TurnOff</rel>
            </op>
            <who>trigger</who>
            <device>TV</device>
         </TriggerAction>
         <TriggerAction>
            <op>
               <rel>TurnOff</rel>
            </op>
            <who>trigger</who>
            <device>Bedroom_Light</device>
         </TriggerAction>
         <TriggerAction>
            <op>
               <rel>TurnOff</rel>
            </op>
            <who>trigger</who>
            <device>Kitchen_Light</device>
         </TriggerAction>
         <TriggerAction>
            <op>
```

```
                <rel>TurnOff</rel>
            </op>
            <who>trigger</who>
            <device>Livingroom_Light</device>
        </TriggerAction>
    </head>
    <body>
        <And>
            <Atom>
                <op>
                    <rel>==</rel>
                </op>
                <var>env_presence_bathroom</var>
                <value>&quot;0&quot;</value>
            </Atom>
            <Atom>
                <op>
                    <rel>==</rel>
                </op>
                <var>env_presence_livingroom</var>
                <value>&quot;0&quot;</value>
            </Atom>
            <Atom>
                <op>
                    <rel>==</rel>
                </op>
                <var>env_presence_kitchen</var>
                <value>&quot;0&quot;</value>
            </Atom>

            <!-- as many conditions as enviromental presence sensors in the house -->

            <Atom>
                <op>
                    <rel>==</rel>
                </op>
                <var>env_presence_bedroom</var>
                <value>&quot;0&quot;</value>
            </Atom>
        </And>
    </body>
</Implies>
<Properties>
    <occurrence>Continuous</occurrence>
    <enabled>false</enabled>
</Properties>
</Trigger>
```

By relying on the virtual devices described above, the same rule can be re-written as follows:

```
<Trigger>
   <name>Electrical devices OFF</name>
   <Implies>
      <head>
         <TriggerAction>
            <op>
               <rel>TurnOff</rel>
            </op>
            <who>trigger</who>
            <device>ListOfElectricalDevices</device>
         </TriggerAction>
      </head>
      <body>
         <And>
            <Atom>
               <op>
                  <rel>==</rel>
               </op>
               <var>env_presence</var>
               <value>&quot;Out of home&quot;</value>
            </Atom>
         </And>
      </body>
   </Implies>
   <Properties>
      <occurrence>Continuous</occurrence>
      <enabled>false</enabled>
   </Properties>
</Trigger>
```

5.2 Rule Definition

In the current implementation of the BackHome project, some proactive context-trigger actions have been designed and developed. Context-Trigger actions are clear examples of the proactive nature of AmI: whenever a rule condition is met the corresponding action is triggered.

5.2.1 Automatic Rule Definition

Rules can be automatically generated depending on the context. Thus, the TMHSS collects information about the activity of the user and learns users habits. For instance, let us consider a user that every night at 9.00 PM watches the TV when s/he is staying at home. Studying users habits, the system learns this context and assumes that at around 9.00 PM the user may turn ON the TV. Hence, the system automatically generates a rule that displays a TV shortcut in the BCI matrix (i.e., the user interface). The corresponding rule is the following:

```
<Trigger>
   <name>Variable Shortcut</name>
   <Implies>
      <head>
         <TriggerAction>
            <op>
               <rel>Shortcut_Turn_On_TV </rel>
            </op>
            <who>trigger</who>
            <device>TV</device>
         </TriggerAction>
      </head>
      <body>
         <And>
            <Atom>
               <op>
                  <rel>==</rel>
               </op>
                <var>env_presence</var>
               <value>&quot;Livingroom&quot;</value>
            </Atom>
            <Atom>
               <op>
                  <rel>&gt;</rel>
               </op>
               <var>time_hourOfDay</var>
               <value>21</value>
            </Atom>
            <Atom>
               <op>
                  <rel>==</rel>
               </op>
               <var>TV_Status</var>
               <value>Off</value>
            </Atom>
         </And>
      </body>
   </Implies>
   <Properties>
      <occurrence>Continuous</occurrence>
      <enabled>false</enabled>
   </Properties>
</Trigger>
```

5.2.2 Rule Definition by the Caregiver

In BackHome, ATML rules are configurable by the end-user through a dedicated interface. In particular, since the end-user is not able to interact directly with the system, a suitable user interface for the caregivers has been provided at the user station. This interface facilitates the creation of rules by the caregiver without requiring any ATML knowledge.

Figure 4 shows the caregiver interface, in which five rules have been defined. It is easy to note that the "Lights OFF" rule in the Figure is aimed at turning off all the home lights after the 9 PM, if the user is at the bedroom and some lights are turned on.

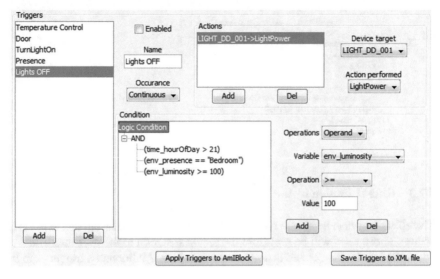

Fig. 4 Trigger definition caregiver interface

The corresponding—automatically generated—ATML translation is the following:

```
<Trigger>
     <name>Lights OFF</name>
     <Implies>
        <head>
           <TriggerAction>
              <op>
                 <rel>LightPower</rel>
              </op>
              <who>trigger</who>
              <device>LIGHT_DD_001</device>
           </TriggerAction>
        </head>
        <body>
           <And>
              <Atom>
                 <op>
                    <rel>&gt;</rel>
                 </op>
                 <var>time_hourOfDay</var>
                 <value>21</value>
              </Atom>
              <Atom>
                 <op>
                    <rel>==</rel>
                 </op>
                 <var>env_presence</var>
                 <value>&quot;Bedroom&quot;</value>
              </Atom>
              <Atom>
                 <op>
                    <rel>&gt;=</rel>
                 </op>
```

```
              <var>env_luminosity</var>
              <value>100</value>
          </Atom>
        </And>
      </body>
    </Implies>
    <Properties>
        <occurrence>Continuous</occurrence>
        <enabled>false</enabled>
    </Properties>
</Trigger>
```

5.2.3   Rule Definition by the Therapist

Therapists may also be able to remotely define ATML rules to be applied at the user station. Those rules will be sent from the therapist station to the user station and—when triggered—the defined action will be executed. Additionally, triggers can be

Fig. 5 Trigger definition therapist interface

configured to send alerts to therapists to take actions based on the corresponding conditions.

Similarly to caregivers, therapists can define rules through a dedicated user interface, integrated in the therapist station. As shown in Fig. 5, the therapist may define a personalized set of user-specific rules that include the conditions and the corresponding actions. corresponding—automatically generated—ATML translation is the following:

```
<Trigger>
      <name>Trigger</name>
      <Implies>
         <head>
            <TriggerAction>
               <op>
                  <rel>Send Alerts to Therapist</rel>
               </op>
               <who>trigger</who>
               <device>Alert</device>
            </TriggerAction>
         </head>
         <body>
            <And>
               <Atom>
                  <op>
                     <rel>==</rel>
                  </op>
                  <var>Room</var>
                  <value>&quot;Bedroom&quot;</value>
               </Atom>
               <Atom>
                  <op>
                     <rel>&gt;</rel>
                  </op>
                  <var>hours in bed</var>
                  <value>10</value>
               </Atom>
            </And>
         </body>
      </Implies>
      <Properties>
         <occurrence>Continuous</occurrence>
         <enabled>true</enabled>
      </Properties>
</Trigger>
```

Remotely managing rule definitions involves a handshaking between user station and therapist station, depicted in Fig. 6. Firstly, the therapist station needs to acquire the definition of the available variables handled by the user station and their allowed values in order to render the interface to the therapist. Secondly, the therapist station, once the therapist has set the appropriate rules, must construct an ATML file which is to be sent to the user station, the actor in charge of handling those rules. Lastly, if the trigger outcome implies sending an alert to the therapist, the user station must be able to send the appropriate information to the therapist station to show the trigger outcomes.

In contrast with the rule definition done by the caregiver in the user station where ATML files are generated by the user station itself, the definition of rules by therapists

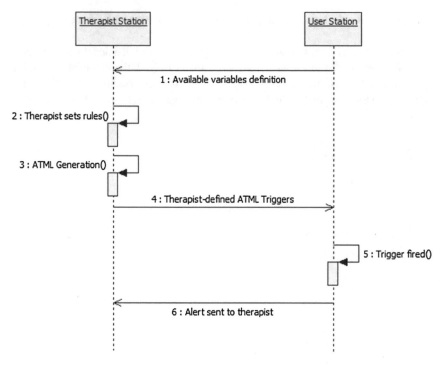

Fig. 6 Therapist rule definition message handshaking

involves the generation of the ATML-compliant file by the therapist station. This use case showcases ATML portability, being also used as a model for distributing remotely generated rules.

6   A Real Scenario

To better illustrate the objectives of the overall project and the central role of AmI, let us illustrate a reference real scenario.[1]

Paul is a 30-year old man depressed about his recent stroke. He used to be a successful painter, but since he became tetraplegic, he has not been able to paint. After staying a long time in hospital, now is the time for Paul to go back home.

Although Paul does not want to try new technologies, Dr. Jones suggests him to try to use a BNCI system at home, because he heard good things about the new BackHome system. After a few weeks, Paul becomes more motivated and returns to painting by using Brain Painting. Dr. Jones notes the progression in Paul's daily activities and in his mood, and the corresponding general improved quality of life.

---

[1]Names have been changed for privacy reasons.

During his normal day, Paul performs several activities. First, Paul's sister helps him wearing the BCI cap that monitors his cerebral activity. Thanks to it, Paul is able to autonomously accomplish goals that are otherwise impossible without the need of a carer.

Through the adoption of the BCI cap, Paul is able to control elements of his home by himself, such as the TV, and IP camera, and switch the lights on and off. Paul also feels more connected to the world because he can surf the Internet, check his email and communicate with relatives, colleagues and friends via Facebook and Twitter.

Paul can also perform CR tasks by playing games recommended by the therapists, according to the level of disability and his needs.

When he wants to have a break, Paul likes watching movies. He is able to autonomously start the multimedia player, select a movie from the menu and play it.

6.1   User Interface Adaptation

Paul is in his living room and he wants to turn on the TV. To do so, normally, Paul has to interact with the BCI matrix, select the option "Smart Home" and then the TV icon and, eventually, turn it on with another command (see Fig. 7).

Thanks to the context-aware system, Paul's position is known and a shortcut in the matrix is automatically displayed with the TV commands. The same can be done knowing that every day at 9 PM, Paul turns on the TV to watch his favorite program. The corresponding user interface with that shortcut is shown in Fig. 8.

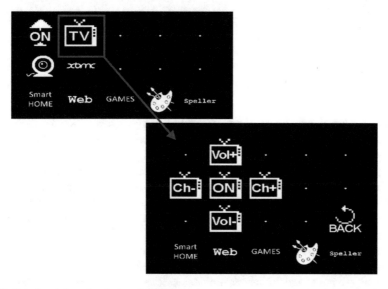

Fig. 7 Interface interaction to turn the TV on

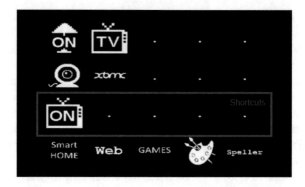

Fig. 8 Interface interaction to turn the TV on when the corresponding shortcut is suggested by the system

Conditions

⌐ Logic Condition
 ⌐ AND
  ⊢ Room equals Bedroom
  ⌐ Hours in bed greater than 10

Operations   Operand

Variable   hours in bed

Operation   lower than

Value   3

Actions

Alert -> Send Alert to Therapist

Device target   Alert

Action performed   Send Alert to all

Add

Remove

Fig. 9 Therapist rule definition detail

6.2 Sending Alerts to Therapist

Dr. Jones fears of Paul being ill and would like to check that he keeps healthy. For such purpose, he logs into the therapist station and goes into the triggers functionality to set rules to be notified when something abnormal is detected.

In particular, Dr. Jones is interested in tracking the sleeping pattern. In fact, since Paul has a very strong disability, he may feel depressed. Detecting how much time the user is sleeping can help Dr. Jones detect a depression at an early stage of the disease. To this end he sets the following rule to raise an alarm if he spends more than ten hours at bed in a day as the Fig. 9 shows.

The system starts the process described and after a week controlling the sleep pattern of Paul the system detects an abnormal situation. Paul has been at bed for more than ten hours and the user station detect it and rises an alert to Dr. Jones. He

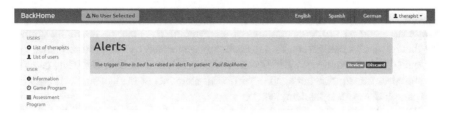

Fig. 10 Alert message at the therapist station when an alert arrives

receives a message at the dashboard of the system showing the alert that has been triggered and the user who is involved (an example is given in Fig. 10).

After receiving the alert, Dr. Jones contacts with Paul's caregiver in order to know the reason of this situation. Finally, Paul has got the flu and is already under treatment and Dr. Jones dismisses the alert. The therapist can also actuate in consequence alerting the caregiver to take special care or changing the therapy to avoid this kind of situations.

7 Conclusions

The application of ambient intelligence to assistive technologies and remote monitoring systems, particularly on healthcare and assistance to dependent people at their homes, opens new possibilities in improving the efficiency and functionality of these systems. The information collected through automated sensors usually creates large volumes of uncertain or incomplete data, making it difficult to exploit it to the benefit of the users (patients, care givers, and health professionals). Defining how to interpret the data and deciding which actions should be taken based on such data becomes a challenging task in those scenarios.

BackHome aims to study the transition from the hospital to the home for people with disabilities, focusing on how they use a Brain/Neural Computer Interface in both settings, with the objective of improving users' quality of life. This project proposes a telemonitoring and home support system able to monitor the evolution of the user's daily life activity, through an array of sensors, and to assist the user, through a set of actuators.

The telemonitoring and home support system uses a rule-based approach to handle efficiently the system adaptation, personalization, alarm triggering, and control over the environment. By defining a suitable set of rules, the system can be customized to the needs of the users. The generation of these rules is key to the correct response of the system, and therefore several methods to define them are needed to be used independently. To facilitate the rule definition process, a language has been defined, namely ATML. This language establishes a platform-independent and flexible definition of rules that allow to describe the relationship between specific conditions and trigger actions.

The use of this language makes possible to create or modify rules by different means, and by different modules of the system. They can be manually defined by therapists and caregivers, or automatically generated by the system taking into account dynamic context conditions and user's habits and preferences. They can also be stored or transmitted between the different elements of the system, as it uses is a formal, platform-independent, syntax.

To ease the manual creation of rules, the concept of "virtual device" is introduced. Virtual devices are software elements that mash together information from two or more sensors in order to make some inference and provide new information. This creates an abstraction that can be used as a regular sensor, effectively distributing the logic of the system. By using virtual devices, the manual generation of rules is greatly simplified. A caregiver can create natural rules that affect several sensors without having to consider each sensor individually. The abstraction of sensor information creates a powerful layered system, which provides a simpler, more readable and more natural description of the rules. Virtual devices can be also useful to automatic systems based on machine-learning, simplifying the calculation of features for the algorithm to train on.

In conclusion, with the definition of a formal language for rule definition, ATML, as well as the use of virtual devices, the BackHome project has achieved to create a home assistance and monitoring system that is powerful yet simple to configure and adapt. Thanks to the proposed telemonitoring and home support system, end-users can rely, for instance, on the automatic customization of the BNCI interface based on their context and caregivers can adjust without difficulties the system to the user's need.

Acknowledgements The research leading to these results has received funding from the European Community's, Seventh Framework Programme FP7/2007-2013, BackHome project grant agreement n. 288566.

References

1. Aarts, E., Harwig, R., Schuurmans, M.: The invisible future. chap. Ambient Intelligence, pp. 235–250. McGraw-Hill, Inc., New York, NY, USA (2002). http://dl.acm.org/citation.cfm?id=504949.504964
2. Aarts, E., de Ruyter, B.: New research perspectives on ambient intelligence. J. Ambient Intell. Smart Environ. 1(1), 5–14 (2009)
3. Acampora, G., Loia, V.: Using fml and fuzzy technology in adaptive ambient intelligence environments. Int. J. Comput. Intell. Res. 1(1), 171–182 (2005)
4. Acampora, G., Loia, V.: Using fuzzy technology in ambient intelligence environments. In: The 14th IEEE International Conference on Fuzzy Systems, 2005. FUZZ'05, pp. 465–470. IEEE (2005)
5. Augusto, J.C., Nakashima, H., Aghajan, H.: Ambient intelligence and smart environments: a state of the art. In: in Handbook of Ambient Intelligence and Smart Environments, pp. 3–31. Springer (2010)

6. Barger, T., Brown, D., Alwan, M.: Health status monitoring through analysis of behavioral patterns. In: 8th congress of the Italian Association for Artificial Intelligence (AI*IA) on Ambient Intelligence, pp. 22–27. Springer (2003)
7. Barger, T.S., Brown, D.E., Alwan, M.: Health-status monitoring through analysis of behavioral patterns. IEEE Trans. Syst. Man Cybern. Part A: Syst. Hum. **35**(1), 22–27 (2005)
8. Brown, P., Bovey, J., Chen, X.: Context-aware applications: from the laboratory to the market-place. Pers. Commun. IEEE **4**(5), 58–64 (1997)
9. Brown, P.J.: The Stick-e document: a framework for creating context-aware applications. In: Proceedings of EP'96, Palo Alto, pp. 259–272 (1996)
10. Carneiro, D., Costa, R., Novais, P., Machado, J., Neves, J.: Simulating and monitoring ambient assisted living. In: Proceedings of the ESM (2008)
11. Casale, P., Fernández, J.M., Rafael, X., Torrellas, S., Ratsgoo, M., Miralles, F.: Enhanching user experience with brain neural computer interfaces in smart home environments. In: 8th IEEE International Conference of Intelligent Environments 2012, INTENV12 (2012)
12. Celler, B., Ilsar, E.D., Earnshaw, W.: Preliminary results of a pilot project on remote monitoring of functional health status in the home. In: Engineering in Medicine and Biology Society, 1996. Bridging Disciplines for Biomedicine. Proceedings of the 18th Annual International Conference of the IEEE, vol. 1, pp. 63–64 (1996)
13. Chan, M., Hariton, C., Ringeard, P., Campo, E.: Smart house automation system for the elderly and the disabled. In: IEEE International Conference on Systems, Man and Cybernetics, 1995. Intelligent Systems for the 21st Century, vol. 2, pp. 1586–1589 (1995)
14. Cook, D.J., Augusto, J.C., Jakkula, V.R.: Ambient Intelligence: Technologies, Applications, and Opportunities (2007)
15. Cook, D.J., Youngblood, G.M., Jain, G.: Algorithms for smart spaces. Computer and Engineering for Design and Applications, Wiley, Technology for Aging, Disability and Independence (2008)
16. Corchado, J., Bajo, J., Tapia, D., Abraham, A.: Using heterogeneous wireless sensor networks in a telemonitoring system for healthcare. IEEE Trans. Inf. Technol. Biomed. **14**(2), 234–240 (2010)
17. Daly, J., Armstrong, E., Miralles, F., Vargiu, E., Müller-Putz, G., Hintermller, C., Guger, C., Kuebler, A., Martin, S.: Backhome: brain-neural-computer interfaces on track to home. In: RAatE 2012—Recent Advances in Assistive Technology & Engineering (2012)
18. Dey, A.K.: Understanding and using context. Pers. Ubiquitous Comput. **5**(1), 4–7 (2001)
19. Farringdon, J., Nashold, S.: Continuous body monitoring. In: Cai, Y. (ed.) Ambient Intelligence for Scientific Discovery. Lecture Notes in Computer Science, vol. 3345, pp. 202–223. Springer, Berlin (2005)
20. Fernández, J.M., Dauwalder, S., Torrellas, S., Faller, J., Scherer, R., Omedas, P., Verschure, P., Espinosa, A., Guger, C., Carmichael, C., Costa, U., Opisso, E., Tormos, J., Miralles, F.: Connecting the disabled to their physical and social world: the BrainAble experience. In: TOBI Workshop IV Practical Brain-Computer Interfaces for End-Users: Progress and Challenges (2013)
21. Fernández, J.M., Torrellas, S., Dauwalder, S., Solà, M., Vargui, E., Miralles, F.: Ambient-intelligence trigger markup language: a new approach to ambient intelligence rule definition. In: 13th Conference of the Italian Association for Artificial Intelligence (AI*IA 2013). CEUR Workshop Proceedings, vol. 1109 (2013)
22. Fuentes, L., Jimenez, D., Pinto, M.: An ambient intelligent language for dynamic adaptation. In: Proceedings of Object Technology for Ambient Intelligence workshop (OT4AmI), Glasgow, Uk (2005)
23. Grosof, B.N.: Representing e-commerce rules via situated courteous logic programs in RuleML (2003)
24. Münssinger, J., Halder, S., Kleih, S., Furdea, A., Raco, V., Hösle, A., Kübler, A.: Brain painting: first evaluation of a new braincomputer interface application with als-patients and healthy volunteers. Front Neurosci. **4**, 182 (2010). doi:10.3389/fnins.2010.00182

25. Nambu, M., Nakajima, K., Noshiro, M., Tamura, T.: An algorithm for the automatic detection of health conditions. Eng. Med. Biol. Mag. IEEE **24**(4), 38–42 (2005)
26. Navarro, A., Ceccaroni, L., F., V., Torrellas, S., Miralles, F., Allison, B., Scherer, R., Faller, J.: Context-awareness as an enhancement of brain-computer interfaces. In: III International Workshop on Ambient Assisted Living IWAAL 2011 (2011)
27. Ogawa, M., Ochiai, S., Shoji, K., Nishihara, M., Togawa, T.: An attempt of monitoring daily activities at home. In: Engineering in Medicine and Biology Society, 2000. In: Proceedings of the 22nd Annual International Conference of the IEEE, vol. 1, pp. 786–788 (2000)
28. Ogawa, M., Suzuki, R., Otake, S., Izutsu, T., Iwaya, T., Togawa, T.: Long term remote behavioral monitoring of elderly by using sensors installed in ordinary houses. In: 2nd Annual International IEEE-EMB Special Topic Conference on Microtechnologies in Medicine & Biology, pp. 322–325. IEEE (2002)
29. Papamarkos, G., Poulovassilis, A., Wood, P.T.: Event-Condition-Action Rule Languages for the Semantic Web. In: Proceedings of Workshop on Semantic Web and Databases, Palo Alto, pp. 309–327 (2003)
30. Schmidt, A.: Implicit human computer interaction through context. Pers. Technol. **4**(2–3), 191–199 (2000)
31. Tabar, A.M., Keshavarz, A., Aghajan, H.: Smart home care network using sensor fusion and distributed vision-based reasoning. In: Proceedings of the 4th ACM International Workshop on Video Surveillance and Sensor Networks, VSSN '06, pp. 145–154. ACM, New York, NY, USA (2006). doi:10.1145/1178782.1178804
32. Tapia, D.I., Corchado, J.M.: An ambient intelligence based multi-agent system for alzheimer health care. Int. J. Ambient Comput. Intell. (IJACI) **1**(1), 15–26 (2009)
33. Vargiu, E., Fernández, J.M., Miralles, F.: Context-aware based quality of life telemonitoring. In: Lai, C., Giuliani, A., Semeraro, G. (eds.) Distributed Systems and Applications of Information Filtering and Retrieval. DART 2012: Revised and Invited Papers. (inpress)
34. Vargiu, E., Fernández, J.M., Torrellas, S., Dauwalder, S., Solà, M., Miralles, F.: A sensor-based telemonitoring and home support system to improve quality of life through bnci. In: 12th European AAATE Conference (2013)
35. Yamaguchi, A., Ogawa, M., Tamura, T., Togawa, T.: Monitoring behavior in the home using positioning sensors. In: Proceedings of the 20th Annual International Conference of the IEEE Engineering in Medicine and Biology Society, 1998, vol. 4, pp. 1977–1979. IEEE (1998)
36. Youngblood, G.M., Cook, D.J., Holder, L.B.: A learning architecture for automating the intelligent environment. In: Proceedings of the 17th conference on Innovative applications of artificial intelligence—Volume 3, IAAI'05, pp. 1576–1581. AAAI Press (2005)
37. Zadeh, L.A.: Fuzzy sets. Inf. Control **8**(3), 338–353 (1965)
38. Zhou, F., Jiao, J., Chen, S., Zhang, D.: A case-driven ambient intelligence system for elderly in-home assistance applications. IEEE Trans. Sys. Man, Cybern. Part C: Appl. Rev. **41**(2), 179–189 (2011)

Dense Semantic Graph and Its Application in Single Document Summarisation

Monika Joshi, Hui Wang and Sally McClean

Abstract Semantic graph representation of text is an important part of natural language processing applications such as text summarisation. We have studied two ways of constructing the semantic graph of a document from dependency parsing of its sentences. The first graph is derived from the subject-object-verb representation of sentence, and the second graph is derived from considering more dependency relations in the sentence by a shortest distance dependency path calculation, resulting in a dense semantic graph. We have shown through experiments that dense semantic graphs gives better performance in semantic graph based unsupervised extractive text summarisation.

Keywords Semantic graph · Dense semantic graph
Single document summarisation

1 Introduction

Information can be categorized into many forms—numerical, visual, text, and audio. Text is abundantly present in online resources. Online blogs, Wikipedia knowledge base, patent documents and customer reviews are potential information sources for different user requirements. One of these requirements is to present a short summary of the originally larger document. The summary is expected to include important information from the original text documents. This is usually achieved by keeping the informative parts of the document and reducing repetitive information. There are two types of text summarization: multiple document summarisation and single document

M. Joshi (✉) · H. Wang
University of Ulster, BT37 0QB, Co. Antrim, UK
e-mail: joshi-m@email.ulster.ac.uk

H. Wang
e-mail: H.Wang@ulster.ac.uk

S. McClean
University of Ulster, BT52 1SA, Co. Londonderry, UK
e-mail: sally@infc.ulst.ac.uk

© Springer International Publishing AG 2018
C. Lai et al. (eds.), *Emerging Ideas on Information Filtering and Retrieval*,
Studies in Computational Intelligence 746, https://doi.org/10.1007/978-3-319-68392-8_4

summarization. The former is aimed at removing repetitive content in a collection of documents. The latter is aimed at shortening a single document whilst keeping the important information. Single document summarisation is particularly useful because large documents are common especially in the digital age, and shortening them without losing important information is certain to save time for the users/readers. The focus of our research is on single document summarisation. In order to process a text document, it should be broken down into parts and then represented in a suitable form to facilitate analysis. Various text representation schemes have been studied, including n-gram, bag of words, and graphs. In our research we use graphs to represent a text document. The graph is constructed by utilising semantic relations such as dependency relations between words within the sentence.

We propose a novel graph generation approach, which is an extension of an existing semantic graph generation approach [5] by including more dependencies from dependency parsing of the sentence. This results in dense semantic graph. We evaluated both graphs in a text summarisation task through experiments. Results show that our dense semantic graph outperformed the original semantic graph for unsupervised extractive text summarization. The next section gives a short literature review of the earlier graph based approaches to text summarisation. In Sect. 3, a detailed description is provided concerning the construction of two different semantic graphs that were used in our study. Section 4 discusses extractive summarisation based on these semantic graphs and Sect. 5 describes the experiments and results. After that conclusion of the analysis follows.

2 Previous Work on Graph Based Text Summarisation

Earlier researchers have used graph representation of documents and properties of graphs to extract important sentences from documents to create a short summary. Graph based text summarisation methods such as LexRank [2], TextRank [7] and Opinosis [3] have shown good performance. There are two types of graph that are constructed and used to represent text. Lexical graph uses the lexical properties of text to construct a graph. LexRank and Text Rank are lexical graph based approaches. They construct graphs by connecting two sentences/smaller text units as nodes in the graph based on the degree of content overlap between them.

On the other hand, semantic graph is based on semantic properties of text. Semantic properties are: Ontological relationship between two words such as synonymy, hyponymy; relationship among set of words representing the syntactic structure of sentence such as dependency tree and syntactic trees. A set of words along with the way they are arranged provides meaning. The same set of words connected in different ways gives different meaning.

According to the semantic properties utilised for graph construction, various representations have been reported in literature for semantic graphs [5, 9]. Some of the approaches utilize the lexical database WordNet to generate ontological relations based semantic graph. In this sentences are broken into terms, mapped to WordNet synsets and connected over WordNet relations [8]. In one of the approaches called semantic Rank [11], sentences are connected as nodes and the weight of the edges between them is the similarity score calculated by WordNet and Wikipedia based similarity measures. Other approaches to generate semantic graphs try to utilize the dependency relations of words in a sentence along with the ontological relations between words. Utilizing this particular order of connection also forms the basis of research work done on semantic graphs in our study. In this area of semantic graph generation most of the work has been concentrated on identifying logical triples (subject-object-predicate) from a document and then connecting these triples based on various semantic similarity measures [5]. Predicate (or verb) is the central part of any sentence, which signifies the main event happening within the sentence. Thus it was mostly agreed to consider the verb and its main arguments (subject and object) as the main information presented in the sentence, and use this as a basic semantic unit of the semantic graph. Various researches have been done on this graph in the field of supervised text summarisation.

We have evaluated two semantic graphs which are based on the dependency structure of words in a sentence. The first graph is triple (subject-object-verb) based semantic graph proposed by Leskovec et al. [5]. The second graph is a novel approach of semantic graph generation proposed in this paper, based on the dependency path length between nodes. Our hypothesis is that moving to a dense semantic graph, as we have defined it, is worthwhile. The principle idea behind this new graph has been used in earlier research in kernel based relation identification [1]. However it has not been used for construction of a semantic graph for the complete document. The next section describes more details about this graph.

3   Semantic Graphs

In the research carried out in this paper, we have analysed the difference between performances when more dependency relations than just subject-object-verb are considered to construct a semantic graph of the document. In this direction, we have developed a methodology to select the dependencies and nodes within a shortest distance path of dependency tree to construct the semantic graph. First we will describe the previous use of graphs and then we will introduce the graph generated by our methodology.

3.1   Semantic Graph Derived from a Triplet (Subject-Object-Verb)

Leskovec et al. [5] has described this graph generation approach for their supervised text summarization, where they train a classifier to learn the important relations between the semantic graph of a summary and the semantic graph of an original text document. In this graph the basic text unit is a triple extracted from sentence: subject-verb-object. This is called triple as there are three connected nodes. Information such as adjectives of subject/object nodes and prepositional information (time, location) are kept as extra information within the nodes. After extracting triples from every sentence of the text document two further steps are taken: (i) co-reference and anaphora resolution: all references to named entities (Person, Location etc.) and pronoun refeences are resolved. (ii) Triples are connected if their subject or object nodes are synoymous or referring to the same named entity. Thus a connected semantic graph is generated.

3.2   Dense Semantic Graphs Generated from Shortest Dependency Paths Between Nouns/Adjectives

We have observed that various named entities such as location/time which are important information, are not covered in the subject-predicate-object relations. As this information is often added through prepositional dependency relations, it gets added to nodes as extra information in the semantic graph generated by previous approaches. However these named entities hold significant information to influence ranking of the sentences for summary generation and to connect nodes in the semantic graph. This has formed the basis of our research into new way of semantic graph generation. First we elaborate the gaps observed in previous approach of semantic graph generation and then give the details of the new semantic graph.

3.2.1   Gaps Identified in Triple (subject-Object-Verb) Based Semantic Graph

The kind of information loss observed in the previous semantic graphs has been de-scribed below:

- Loss of links between words in sentence

  Some connections between named entities are not considered because they do not come into the subject/object category. This information is associated with subject/object, but does not get connected in the semantic graph, as they are not directly linked through a predicate. For example consider the sentence below:

  President Obama's arrival in London created a joyful atmosphere.
  The triple extracted from this sentence is:
  Arrival→ create→ *atmosphere*

  Here the information *London*, *Obama* is added as extra information to node *Arrival*, and *Joyful* is added to node *Atmosphere*. However a direct link between *London* and *atmosphere* is missing, whereas a reader can clearly see this is atmosphere of London. This connection can be identified in our shortest dependency path graph as shown below:

London-prep-in → *Arrival-nsubj*→ *created-dobj* → *atmosphere*

- Loss of inter-sentence links between words

Some named entities which are not subject/object in one sentence are subject/object of another sentence. When creating a semantic graph of complete document, these entities are the connecting words between these sentences. In the previous graph these connections are lost as shown below by two sentences.

He went to church in Long valley.
One of the explosions happened in Long Valley.
The triple extracted from these sentences is:
He→went→church
Explosion→happened→long valley

In the semantic graph derived from triples of the above 2 sentences, we do not have both these sentences connected, because the common link *Long Valley* is hidden as extra information in one semantic graph.

- Identification of subject is not clear

In a few cases, identification of a subject for the predicate is not very accurate with current dependency parsers. This case occurs in the clausal complement of verb phrase or adjectival phrases called dependency relation 'xcomp'. Here the determination of subject for clausal complement is not very accurate, as the subject is external.

3.2.2 Construction of Shortest Distance Dependency Path Based Semantic Graph

To overcome these gaps, we construct the semantic graph by connecting all noun and adjectives which are connected within a shortest path distance in the dependency tree of that sentence. From the literature review it has been identified that nouns are the most important entities to be considered for ranking sentences. So we have decided to include nouns as nodes in the semantic graph. We also considered adjectives, as they modify nouns and may present significant information. The length of the shortest path is varied from 2–5 to analyse its effect on the efficiency of the PageRank score calculation. The following steps are followed to construct the semantic graph

- Co-reference resolution of named entities
  The text document is preprocessed to resolve all co-references of named entities. We replace the references with the main named Entity for Person, Location, and organization.

Fig. 1 The triple based
semantic graph for the text
excerpt taken from DUC
2002 data

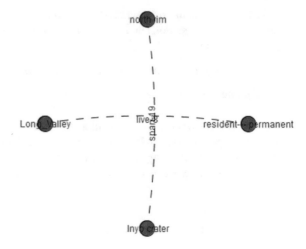

- Pronominal resolution

  After co-reference resolution, text is preprocessed for pronominal resolution. All references (he, she, it, who) are resolved to referring named entities and get replaced in text.

- Identifying nodes and edges of the semantic graph. The shortest path distance based Semantic graph is defined as G = (V, E), Where

$$V = \cup_{word_i \in document} Word_i : pos(Word_i) \in \{JJ*, NN*\} \tag{1}$$

In (1) $pos(Word_i)$ provides part of the speech tag of $Word_i$. According to Penn tag set for part of speech tags, "JJ" signifies Adjectives and "NN" signifies Noun.

$$Edgeset\, E = \cup_{(u,v \in V)} (u, v) : SD(u, v) \leq limit \tag{2}$$

In (2) $SD(u, v)$ is the shortest distance from u to v in the dependency tree of that sentence and $limit$ is the maximum allowed shortest path distance, which is varied from 2–5 in our experiments.

We have used Stanford CoreNLP package for co-reference resolution, identification of named entities and dependency parse tree generation [4, 10]. To develop the graphs and calculate the page rank scores of nodes we use the JUNG software package.[1] First we extract dependency relations for each sentence. Then we generate a temporary graph for the dependency tree of that sentence in JUNG. Then Dijkstra's shortest path algorithm is applied to find the shortest distance between nodes. From this temporary graph we find vertices and edges based on Eqs. (1) and (2) to construct the semantic graph.

---

[1]https://www.jung.sourceforge.net/.

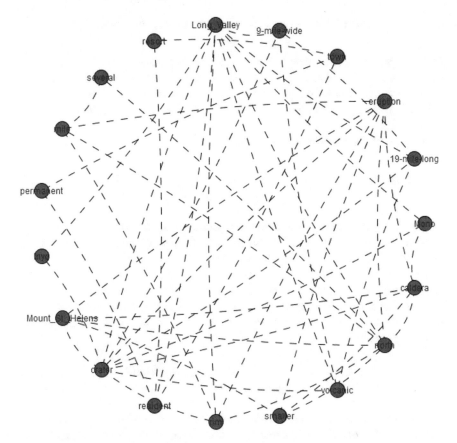

Fig. 2 Sematic graph based on the shortest dependency path between nouns/adjectives (shortest distance = 2) for the text excerpt taken from the DUC 2002 data

Figures 1 and 2 show two graphs, triple based semantic graph and shortest distance dependency path based semantic graphs for the given excerpt of 2 sentences below, taken from the Long Valley document of DUC2002 data.

A text excerpt taken from DUC 2002 data.

The resort town's 4,700 permanent residents live in Long Valley, a 19-mile-long, 9-mile-wide volcanic crater known as a caldera. Eruptions somewhat smaller than Mount St. Helens' happened 550 years ago at the Inyo craters, which span Long Valley's north rim, and 650 years ago at the Mono craters, several miles north of the caldera.

The next section describes the methodology to rank sentences based on the semantic graph described in this section.

4   Extraction of Sentences for Summary Generation

In this paper we want to analyse the impact of dense semantic graphs on text summarisation and provide a comparison with the summarisation results of earlier triple based semantic graphs. To achieve this, first we rank the semantic graph by one of the graph ranking algorithm. We have used PageRank method to rank the semantic graph nodes.

The PageRank score of $node_i$ is calculated as:

$$PageRank(node_i) = (1 - d) + d * \sum_{node_j \in In(node_i)} \frac{PageRank(node_j)}{Out(node_j)} \qquad (3)$$

where d is the probability of jumping from $node_i$ to any random node in the graph, typically set between 0.1 and 0.2. $In(node_i)$ is the set of incoming edges to $node_i$ and $Out(node_j)$ is the set of outgoing edges of $node_j$. Initially PageRank of all nodes is intialised with arbitrary values, as it does not affect the final values after convergence. In this paper semantic graphs are undirected graphs so incoming edges of a node are equal to outgoing edges. After calculating PageRank score of the nodes in the semantic graph, the score of sentence S_i in the text document is calculated by following equation:

$$Score_{(S_i)} = \sum_{(node_j \in graph \cap S_i)} PageRank(node_j) \qquad (4)$$

where $node_j$ is the stemmed word/phrase in the graph representation. Scores are normalised after dividing by the maximum score of sentences. After calculating normalized scores of all sentences in the text document, sentences are ordered according to their scores. As per the summary length, higher scoring sentences are taken as summary sentences.

In addition to this summary generation method, we have also tried to analyze impact of including additional features together with PageRank scores on semantic graph based text summarisation. This was done in a separate experimental run where we have included sentence position as an additional feature for scoring of sentences. Since the data we have experimented with is news data, a higher score is given to early sentences of the document. So the score of a sentence S_i after including sentence position, i as a feature is given by:

$$newScore_{(S_i)} = \frac{0.1 \times (Count_{sentences} - i)}{Count_{sentences}} + 0.9 \times Score_{S_i} \qquad (5)$$

After calculating the new score of the sentences, higher scoring sentences are extracted as the summary as in previous summarisation method. The next section describes the experimental setup.

5   Experiments

We have experimented on two single document summarisation corpuses from Document Understanding Conference (DUC), DUC-01 and DUC-02. DUC-01 contains 308 text documents and DUC-02 contains 567 text documents. Both sets have 2 human written summaries per document for evaluation purposes. We have used the ROUGE toolkit to evaluate system generated summaries with reference summaries, that are 2 human generated summaries per document [6]. The ROUGE toolkit has been used for DUC evaluations since the year 2004. It is a recall oriented evaluation metric which matches n-grams between a system generated summary and reference summaries.

$$Rouge - N = \frac{\sum_{S \in \{ReferenceSummaries\}} \sum_{gram_n \in S} Count_{match}(gram_n)}{\sum_{S \in \{ReferenceSummaries\}} \sum_{gram_n \in S} Count(gram_n)} \qquad (6)$$

Rouge-1 is 1-gram metric. Rouge-2 is 2-gram metric. Rouge-W is the longest weighted sequence metric, which gives weight to consecutive longest sequence matches.

ROUGE scores were calculated for different summarisation runs on triple based semantic graphs and shortest dependency distance path based semantic graphs. On triple based graphs two summarisation tasks were run for DUC01 and DUC-02 data. The first considered PageRank only and the second used PageRank, sentence position (Triple based, Triple + position). On the Shortest distance dependency path based semantic graph, 6 summarisation tasks were run for both datasets. The first 4 runs are based on PageRank scores alone by varying shortest distance from 2–5: shortest distance 2 (SD-2), shortest distance 3 (SD-3), shortest distance 4 (SD-4) and shortest distance 5 (SD-5). The fifth and sixth run include sentence position as feature with SD-4 and SD-5(SD-4 + position, SD-5+ position). We have also compared our results with the results of the text summarisation software *Open Text Summarizer (OTS)* [12], which is freely available and has been reported to perform best between other available open source summarizers. OTS differs from our approach as it utilizes statistical information of words in the text and uses a language specific lexicon to identify synonyms and cue words to extract the important sentences as summary sentences. It doesn't utilizes the connection between words, which forms the basis of our approach.

6   Results and Analysis

Figure 3 shows the ROUGE-1, ROUGE-2 and ROUGE-W scores for DUC-01 data achieved by different experimental runs described in Sect. 5. The Rouge evaluation setting was a 100 words summary, 95% confidence, stemmed words and no stop words included during summary comparison. In Fig. 3, we have observed that the

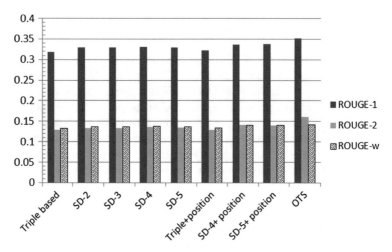

Fig. 3 ROUGE scores obtained for a summarisation test on DUC-01 data

lowest Rouge scores are reported with the triple based experiment. By including position, results for triple based experiment are improved. Rouge-1 scores for SD-2, SD-3, SD-4, and SD-5 improves systematically and are better than triple based and triple based + position. This shows that as the shortest length of dependency path was increased from 2 to 5, the Rouge score has improved due to better ranking of the nodes in the semantic graph. This better ranking can be attributed to more connections found after increasing the path distance to find links in the dependency tree. A similar trend of increase in ROUGE-2 and ROUGE- W scores are observed for experiments on DUC-02 data in SD-2, SD-3, SD-4, SD-5, SD-4+ position, and SD-5+ position. Although benchmark OTS results are always higher than best results achieved by our approach, it is useful to observe that our results are comparable to the benchmark results, as the main purpose of our research is to analyse the impact of dense semantic graphs on text summarisation compared to previous semantic graph.Also as per the documentation of OTS, it is not useful for summarising document where multiple topics are dicussed due to the nature of the alogorithm of OTS, which works on the frequency count of words. But our approach is equally applicable despite of multiple topics as it works on the connection between words not the count of words. Table 1 gives results in a numerical form for the DUC-01 experiments. Figure 4 and Table 2 shows the scores for the DUC 02 dataset. For both corpuses the ROUGE scores improves on shortest dependency based graph, until distance 5. During results analysis we have observed that the ROUGE score decreases or becomes approximately constant if we increase distance after 5.

Including sentence position as a feature, improves the summarisation results on both triple based graph and shortest distance dependency path based semantic graph. Also in this case, ROUGE scores for summarisation run on shortest distance dependency path based semantic graph are higher than for triple based semantic graphs. This also indicates that we can include more features to improve the results further.

Table 1 ROUGE scores for a summarisation test on DUC-01 data

System	Rouge-1	Rouge-2	Rouge-W
Triplet based	0.31793	0.12829	0.13214
SD-2	0.32964	0.13229	0.1354
SD-3	0.3298	0.13301	0.1359
SD-4	0.33037	0.1351	0.13671
SD-5	0.32974	0.13365	0.13621
Triple + position	0.3224	0.12923	0.13355
SD-4+ position	0.33676	0.14106	0.14017
SD-5+ position	0.33753	0.14049	0.14023
OTS	0.35134	0.16039	0.14093

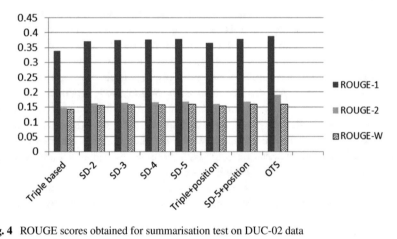

Fig. 4 ROUGE scores obtained for summarisation test on DUC-02 data

Table 2 ROUGE scores for summarisation test on DUC-02 data

System	Rouge-1	Rouge-2	Rouge-W
Triplet based	0.33864	0.14714	0.14143
SD-2	0.37154	0.16221	0.15465
SD-3	0.37494	0.16409	0.1563
SD-4	0.37666	0.16498	0.15694
SD-5	0.37919	0.168	0.15778
Triple + position	0.36465	0.16016	0.15231
SD-4+ position	0.37666	0.16498	0.15694
SD-5+ position	0.37937	0.16846	0.15793
OTS	0. 0.38864	0.18966	0.15766

Overall results indicate that shortest distance based semantic graphs performs better in ranking the sentences and are comparable to benchmark system OTS.

7 Conclusion

PageRank based summarisation is a novel approach for both our approaches. Earlier for triple based semantic graph, PageRank node score was considered as a feature for supervised text summarisation. In this paper we have looked at unsupervised single document summarisation. In the evaluation, we have seen that only PageRank based summarisation results do not exceed the benchmark results, but are comparable. Benchmark OTS system utilises a language specific lexicon for identifying synonymous words and cue terms. In future work, we can include a similar lexicon to identify more relation between words to improve the performance. In this paper we have hypothesised that if more dependency relations are considered for semantic graph generation it gives better PageRank scores and thus improves the ranking accuracy for extraction of summary sentences. Although triple based graphs are more visually understandable they can be enhanced by adding more dependencies. When sentence position was included as an extra feature, it improved the Rouge scores. Also it is noticeable that summarisation results for shortest distance dependency path based semantic graph are similar to results after including the additional feature sentence position. This makes this graph equally useful in domains where sentence position does not have an effect on importance.

In future work we will apply semantic similarity and word sense disambiguation to improve the connectivity of the graph and identify more relations between nodes.

References

1. Bunescu, R.C., Raymond, J.M.: A shortest path dependency kernel for relation extraction. In: Proceedings of the conference on Human Language Technology and Empirical Methods in Natural Language Processing, number October, Vancouver, British Columbia, Canada, 2005, pp. 724–731. Association for Computational Linguistics (2005)
2. Erkan, G., Radev, D.R.: LexRank: graph-based lexical centrality as salience in text summarization. J. Artif. Intell. Res. **22**(1), 457–479 (2004)
3. Ganesan, K., Zhai, C., Han, J.: Opinosis: a graph-based approach to abstractive summarization of highly redundant opinions. In: Proceedings of the 23rd International Conference on Computational Linguistics, number August, Beijing, China, 2010, pp. 340–348. Association for Computational Linguistics (2010)
4. Lee, H., Peirsman, Y., Chang, A., Chambers, N., Surdeanu, M., Jurafsky, D.: Stanford's multi-pass sieve coreference resolution system at the CoNLL-2011 shared task. In: Proceedings of the CoNLL-2011 Shared Task, pp. 28–34. Association for Computational Linguistics, June 2011
5. Leskovec, J., Milic-Frayling, N., Grobelnik, M.: Extracting Summary Sentences Based on the Document Semantic Graph. Microsoft Technical Report TR-2005-07 (2005)

6. Lin, C.-Y., Rey, M., Ouge, R.: A package for automatic evaluation of summaries. In: Proceedings of the ACL-04 Workshop: Text Summarization Branches Out, pp. 74–81, Barcelona, Spain (2004)
7. Mihalcea, R., Tarau, P.: TextRank: bringing order into texts. In: Proceedings of Empirical Methods in Natural Language Processing, Barcelona, Spain (2004)
8. Plaza, L., Díaz, A.: Using semantic graphs and word sense disambiguation techniques to improve text summarization. Procesamiento del Lenguaje Natural Revista **47**, 97–105 (2011)
9. Rusu, D., Fortuna, B., Grobelnik, M., Mladenić, D.: Semantic graphs derived from triplets with application in document summarization. Informatica J. (2009)
10. Toutanova, K., Klein, D., Manning, C.D., Singer, Y.: Feature-rich part-of-speech tagging with a cyclic dependency network. In: Proceedings of the 2003 Conference of the North American Chapter of the Association for Computational Linguistics on Human Language Technology—NAACL '03, vol. 1, pp. 173–180, Morristown, NJ, USA, May 2003. Association for Computational Linguistics (2003)
11. Tsatsaronis, G., Varlamis, I., Nørvåg, K.: SemanticRank: ranking keywords and sentences using semantic graphs. In: COLING'10 Proceedings of the 23rd International Conference on Computational Linguistics, number August, pp. 1074–1082 (2010)
12. Yatsko, V.A., Vishnyakov, T.N.: A method for evaluating modern systems of automatic text summarization. Autom. Doc. Math. Linguist. **41**(3):93–103, June 2007

Web Architecture of a Web Portal for Reliability Diagnosis of Bus Regularity

Benedetto Barabino, Cristian Lai, Roberto Demontis, Sara Mozzoni, Carlino Casari, Antonio Pintus and Proto Tilocca

Abstract In high frequency transit services, bus regularity—i.e., the headway adherence between buses at bus stops—can be used as an indication of service quality, in terms of reliability, by both users and transit agencies. The Web architecture is the entry point of a Decision Support System (DSS), and contains an environment designed for experts in transport domain. The environment is composed of tools developed to automatically handle Automatic Vehicle Location (AVL) raw data for measuring the Level of Service (LoS) of bus regularity at each bus stop and time interval of a transit bus route. The results are represented within easy-to-read control dashboards consisting of tables, charts, and maps, able to perform fast AVL processing and easy accessibility in order to reduce the workload of transit operators. These outcomes show the importance of well-handled and presented AVL data, in order to use them more effectively, improving past analysis done by using, if any, manual methods.

B. Barabino (✉) · S. Mozzoni
Technomobility S.r.l., Cagliari, Italy
e-mail: bbarabino@gmail.com

S. Mozzoni
e-mail: sara.mozzoni@ctmcagliari.it

C. Lai · R. Demontis · C. Casari · A. Pintus
Research and Development in Sardinia, CRS4, Center for Advanced Studies,
Pula, CA, Italy
e-mail: clai@crs4.it

R. Demontis
e-mail: demontis@crs4.it

C. Casari
e-mail: casari@crs4.it

A. Pintus
e-mail: pintux@crs4.it

P. Tilocca
CTM S.p.A., Cagliari, Italy
e-mail: proto.tilocca@ctmcagliari.it

© Springer International Publishing AG 2018   69
C. Lai et al. (eds.), *Emerging Ideas on Information Filtering and Retrieval*,
Studies in Computational Intelligence 746, https://doi.org/10.1007/978-3-319-68392-8_5

Keywords Web-based application · Transit network · Bus regularity
AVL raw data · Service quality

1 Introduction

The measurement of public transport service quality is currently a crucial element
for both users and transit operators [1]. A significant aspect of service quality is
represented by the reliability which can be viewed as the dependability of the transit
service in terms of multidimensional aspects such as waiting and riding times, passenger loads, vehicle quality, safety, amenities and information [2]. Looking at high
frequency bus services—i.e., those in which scheduled temporal headways between
buses are 10/12 min (e.g. [3–6])—regularity is the relevant element of reliability.
This element is addressed in this chapter. In order to avoid misleading information,
the detailed evaluation of regularity, in terms of space and time requires to work on
a great amount of data to be collected and normalized before processing. In addition, for an effective monitoring of the regularity, it is necessary to process data into
user-friendly significant outcomes and to guarantee a fast and pervasive access to
them.

The objective of this work are: (i) to refine, in a Web architecture, the implementation of a methodology to assess the regularity mainly starting with the data
collected by Automatic Vehicle Location (AVL) technology; (ii) to improve the web
architecture specifically designed to support experts in transport engineering domain
for evaluating regularity issues, discussing new elements.

Nowadays, abundance of time-at-location data (i.e., arrival or departure time of a
bus at checkpoints) on buses are collected by AVL technology. These data allow to
perform detailed analysis of the bus service, although their use requires addressing
challenges such as missing data points and possible bus overtaking, and infers data
on original bus schedule and service disruptions (e.g. missed trips). Moreover, while
ad hoc and manual methods of bus operators are thought to operate in data-poor
environments, new and automatic methods must be developed to exploit the rich-data
environments provided by AVL and to handle these data. An efficient management
of the service (i.e., decisions based on data) requires the ability to process data and
quickly present outcomes, which are crucial elements of the most advanced transit
operators. As far as authors know, transit operator executives use spreadsheets to
face this challenge. At first, archived raw AVL data are downloaded in a standard
PC, and separated according to routes. Then, the value of regularity is directly shown
in a separate file or at a later stage calculated by formulas, per route and time period.
Finally, outcomes of the processed data (i.e., the value of regularity) are presented in
a table organized per route direction and time periods. Unfortunately, these activities
are very time- and energy-consuming, because usually performed using procedures
designed to handle few data (i.e., manual methods). Besides, results are available
only locally. Subsequently, the need for fast procedures to effectively process AVL
raw data, quickly present results and guarantee their access from everywhere arises.

As a result, making reference to previous works ([7–9]), we refine in a Web portal the implementation of a method to derive accurate measure of bus regularity -in space i.e., at every bus stop and time i.e., at every time period-, to shed additional light into the diagnosis of service regularity. The method is expected to improve the regularity measurements, which are too often made, when so, at a limited number of checkpoints, on selected routes, and at limited time periods. One more point, based on these measures, transit operator managers will be able to prioritize actions and/or give recommendations to improve the service. In addition, thanks to the possibility to perform fast AVL data processing and thanks to their easy accessibility (provided that a Web connection is available), the workload of transit operators will be reduced.

This chapter is organized as follows. In Sect. 2, we motivate the choice of the regularity indicator, describe the challenges derived from AVL technologies, and mention a number of Web existing tools. In Sect. 3, we briefly discuss the methodology evaluating the regularity. In Sect. 4 the architecture of the system, the used data and its control dashboards are presented. In Sect. 5, we present conclusions and research perspectives.

2 State of the Art

In high frequency services, regularity is a major aspect of service, and a classical topic for the transportation community. The major existing studies in the field, including details on the measure of regularity, AVL technology and existing Web tools, are presented in the following three subsections.

2.1 Measure of Regularity

Bus regularity can be measured by several indicators, which present pros and cons and denote the significant lack of a universal metric [5, 10]. The discussion about the indicators used is not required in this chapter, because already presented in [7, 8]. However, to summarize, we look for an indicator which should satisfy the following features: easy communication (understandable and easy-to-read), objectivity (i.e., without subjective thresholds), customer oriented (penalizing longer waiting times), independence from data distributions and ranking well-established regularity levels. As discussed in [7–9], the Headway Adherence (HA) measured by the Coefficient of Variation of Headways (C_{vh}) proposed by [10] is a good indicator fitting the following requirements:

- although the C_{vh} is not immediately understandable and conveyable, its values represent LoSs ranked in a well-established scale of regularity from A (the best) to F (the worst);
- it is objective; LoS thresholds are related to the probability that a given transit vehicle's actual headway is off headway by more than one-half of the scheduled headway;
- it is customer oriented; every trip is considered in the computation of the C_{vh}, in order to penalize long waiting times at bus stops. The output indicates the probability of encountering an irregular service, even if it is not a measure of severity of the irregularity;
- it does not require particular applicability conditions. Since bus operators sometimes schedule high frequency services irregularly, it is important to consider different headways in different time intervals;
- it can evaluate different regularity conditions and detect bunching phenomena.

Therefore, in this chapter, the Headway Adherence is chosen as a regularity indicator.

2.2   AVL Technology and Regularity

Due to economic constraints and lack of technology, early experiences in the determination of regularity measures were performed at a few random or selected check points of bus route (e.g. [4, 6]). Typically, collected data were aggregated manually in time periods representing slack and peak hours in the morning and in the evening. This way of working generates restricted analysis and leads to limited conclusions. When data are aggregated from checkpoints to route level, one typically loses a considerable amount of information on the regularity between consecutive checkpoints. This procedure is rarely user-oriented, because passengers are mostly concerned with adherence to the headways at their particular bus stop (e.g. [11]). Hence, in order to provide the best possible service to passengers, measures should be performed at every stop of the bus route and for every investigated time period. Therefore, performing regularity measures at all bus stops and time periods removes shortcomings deriving from choosing checkpoints and aggregating data in large time periods. Nowadays, relevant support is provided by AVL technology, because it can collect huge amounts of disaggregated data on different bus stops and time periods. Most importantly, if properly handled and processed, AVL data have the capability to show when and where the service was not provided as planned. However, there are two critical issues which must be faced before being able to perform accurate regularity calculation, otherwise the calculation of regularity will not sufficiently reflect the service that customers experience and this will lead to misleading information. These critical points are:

1. Bus Overtaking (BO) which arises when the succeeding scheduled bus overtakes its predecessor in the route;
2. Missing data point, which consist of Technical Failures (TF) depending on AVL being temporarily out of order, and Incorrect Operations in the Service (IOS), such as missed trips and unexpected breakdowns.

Due to possible BO, buses might not arrive in the right order. For passengers whose aim is to board on the first useful arriving bus at bus stop, BO is irrelevant because the headway is the time elapsed between two consecutive buses, in which the last one may or may not be the scheduled bus. Hence, instead of tracking buses (e.g. [12]), regularity measures should focus on transits (i.e., arrivals or departures) of the first bus arriving at the considered bus stop. TF and IOS result in missing data points, which are not recorded by AVL. Moreover, they result in temporal gaps. Hence, a crucial challenge is to recognize the type of missing data and handle the temporal gaps, because they have a different impact on users. The temporal gaps due to TF lead to an incorrect calculation of headways, because buses actually arrived at bus stop, but they were not recorded by AVL. Considering the temporal gaps due IOS is favorable because they are perceived by users as real. McLeod [13] provided insights, in order to determine temporal gaps due to TF and shown that less than 20% of missing data due to TF leads to good quality headway measures. In [7, 8], in order to recognize and address BO, TF and IOS, a method has been proposed in the case of regularity analysis at the single route and at the whole bus transportation network, respectively. However, in this case two software applications are used to implement the method. Therefore, additional work must be done to implement the method by a single application in order to make AVL data a mainstream source of information when regularity calculation are performed.

2.3 Regularity Web Tools

A key factor for the effective analysis of data is building intelligible performance reports. To date, there are few state of the art of modern Web platforms, specifically designed to provide a Decision Support System (DSS) focused on reliability diagnosis of bus regularity. There are a few research works focusing on Web-based, AVL data visualization, as in [14] or data analysis algorithms and techniques, including a very basic visualization of route paths and speeds using Google maps as in [15]. Conversely, the current state of the art information systems technologies includes mature and reliable tools. Mainstream commercial products, or Web frameworks released under Open Source licenses, are designed and documented to integrate with other systems in order to build complex and large-scale Web applications, usually thanks to the use of Application Programming Interfaces (APIs).

Some noteworthy product and framework categories are: business intelligence (BI) tools, reporting and OLAP systems, as Jaspersoft.[1] or Pentaho[2]; Web architecture development platforms and Content Management Systems (CMS), as the open-source ones like Entando.[3] or Joomla[4]; database management systems (DBMS), like the well-known and broadly adopted MySQL[5] or PostgresSQL.[6]

3 Methodology

In this section we summarize the method implemented in the Web architecture described in Sect. 4. The method is taken from previous author's works ([7, 8]) where further details can be found. The method addresses three main phases, namely: to validate AVL data, to address criticalities in AVL raw data and to determine the value of C_{vh} in order to illustrate the LoS of regularity over space at every bus stops and route direction—and time—at every time period—in a bus transit network.

3.1 The Validation of AVL Raw Data

Specific attention must be paid to bus stops, because bus operators measure regularities at these points, where passengers board and get off. Using this methodology, the relevant elements recorded by AVL at each bus stop for each high frequency route are: day, route, direction, actual and scheduled transit times.

When comparing the numbers of actual and scheduled transits, the lack of data might be observed due to IOS and TF. In this chapter, we contemplate the situation where the transport service is good, according to historical data. As a result, few IOS are expected to occur. Therefore, missing data point are fundamentally TF which must be detected and processed in order to determine correct measures of headways. For this reason, we consider the following three main steps to accept or reject data related to days and months and validate a counting section.

STEP 1. Read daily AVL data at a bus stop of a specific route and check whether the number of recorded transits is larger than or equal to a certain percentage of scheduled transits. This percentage can be set equal to 80% of scheduled transits, because McLeod [13] shown that the estimation of headway variance is still good when 20% of data are missing. If a bus stop meets this criterion in that day, it is used for the next step.

---

[1]http://www.jaspersoft.com.

[2]http://www.pentaho.com.

[3]http://www.entando.com.

[4]http://www.joomla.org.

[5]http://www.mysql.com.

[6]http://www.postgresql.org.

STEP 2. Perform a chi-square test on the set of bus stops selected by STEP 1 to evaluate the approximation of the actual number of transits to scheduled transits. A suitable significance value for this test is $\alpha = 0.05$. If a day satisfies this criterion, it is used for the next step as well as the bus stops of that day.

STEP 3. Collect all bus stops satisfying STEP 2 in a monthly list and compute the ratio between the number of bus stops in the monthly list and the total number of bus stops. If this ratio is larger than a threshold, all monthly data are supposed to be valid. Based on our experience in preliminary tests, we recommend the use of percentages larger than 60%, which is a good threshold value, in order to cover a significant number of bus stops in a route.

A detailed example of steps 1, 2 and 3 is reported in [7].

3.2 The Handling of Criticalities

Data validation is followed by the detection of criticalities in order to correctly calculate headways between buses. This phase is applicable both in case of few and of many unexpected IOS. As illustrated in Sect. 2.2, three types of criticalities might occur: BO, TF and IOS. Since TF and IOS lead to missing data and temporal gaps, they can be addressed together. Gaps must be found and processed by comparing scheduled and actual transit times. Sophisticated AVL databases can be used to match scheduled transit time data (with no gaps) with actual transit time data (with possible gaps). As a result, to address all criticalities, the following steps are carried out:

STEP 4. Address BO by ordering the sequence of actual transit times at bus stops, because BO is irrelevant for the regularity perceived by users, who are not interested in the right schedule of buses.

STEP 5. Fill up tables reporting unpredicted missed trips and unexpected breakdowns. Columns include the day, bus stops, route, direction, scheduled transit times and incorrect operation code, indicating whether it is neither a missed trip or a breakdown.

STEP 6. Consider the table of scheduled service on that day with the attributes of the table in STEP 5, whereas the incorrect operation code will be neglected, because it is an unexpected event.

STEP 7. Match these tables, in order to generate a new table of scheduled transits with a new attribute indicating the incorrect operation code.

STEP 8. Detect TF and IOS. Match the table built in STEP 7 with data at the end of STEP 4 and detect possible gaps, when a transit between two recorded transits is missing.

STEP 9. Correct TF and IOS. Disregard gaps generated by TF, because no real headways can be derived as the difference between two consecutive transits. Keep the gaps generated by IOS, because these gaps are really perceived by users.

A detailed example of steps 4, 8 and 9 is reported in [7, 8].

3.3   The Calculation of Regularity LoS

Given a generic bus stop j at time period t along the direction d of a route r, once data have been validated and criticality have been addressed, we calculate the C_{vh} as follows:

$$C_{vh}^{j,t,d,r} = \frac{\sigma^{j,t,d,r}}{h^{j,t,d,r}} \tag{1}$$

where:

- $\sigma^{j,t,d,r}$ is the standard deviation of the differences between actual and scheduled headway at bus stop j, time interval t, direction d and route r; the values used for the evaluation span over a monthly planning horizon.
- $h^{j,t,d,r}$ is the average scheduled headway at bus stop j, time interval t, direction d and route r.

As shown in Sect. 2.1, the C_{vh} represents a standard indicator used to evaluate bus regularity [10].

In many transit agencies, the standard time interval is one hour. Since transit agencies may add or remove some bus trips to better serve the changing demand, in this chapter $h^{j,t,d,r}$ is computed as the average of headways of scheduled transits times, to account for these additional trips and possible gaps. As illustrated previously, Eq. (1) provides results in a monthly planning horizon whose set is denoted by S. The elementary observation is denoted by $x_i (i = 1, n)$ and represents the precise headway deviation at the end of STEP 9. However, in order to provide monthly aggregated statistics for week and type of day, Eq. (1) is also calculated for them, considering the sub-sets S_1 and S_2. The related elementary observations are denoted by $x_{1j} (j = 1, m)$ and $x_{2k} (k = 1, p)$ and represents the precise headway deviation at the end of STEP 9, when they are related to the week and type of day, respectively. Therefore, to summarize, the results provided by Eq. (1) refer to the considered sets defined as:

- S—the set representing the headway deviations (x_i) within the month.
- S_1—the set representing the headway deviations (x_{1i}) within the considered week in the month.
- S_2—the set representing the headway deviations (x_{2i}) within the considered type of day in the month.

The calculated values of C_{vh} can be converted into the LoS according to [10]. LoSs can be represented by dashboards as illustrated in the next section.

4   Web Architecture

The web architecture is the entry point of the DSS (hereafter the 'system') composed of an environment designed for transit industry experts. Such environment is designed to primarily handle AVL raw data for measuring the LoS of bus regularity at each

Fig. 1 Components of the system

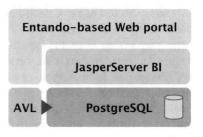

bus stop and time period of a transit bus route. The web architecture is powered by Entando, a Java Open Source portal-like platform for building information, cloud, mobile and social enterprise-class solutions. It natively combines portal, web CMS and framework capabilities. The portal provides dashboards that support experts with significant and useful data. A main feature of the dashboards is to identify where and when regularity problems occur. Dashboards are intended to show summaries and consist of tables, charts and maps. The items located on the dashboards include:

- regularity table—a table showing the LoS; the executive may select a route direction and a time period; the table will show for each bus stop of the selected route the LoS of regularity in different colors;
- regularity line chart—a multi-line chart showing one chart for each time period; the charts are distinguished by color and may be compared directly;
- mapper—a Google Maps technology based interface; the map shows the path of a selected route direction. The bus stops associated with the route are represented with different colors depending on the LoS of regularity; the executive may interact with the map changing the time slot within the time period.

4.1 Architecture of the System

The web architecture is part of the whole system. The Fig. 1 shows the components of the system.

At the bottom, the AVL is the technology which collects raw data during the transport service. As illustrated in Sect. 3, collected data deals with real measures including actual and scheduled bus transit time, bus information, route information, bus stops information. In case of non-homogeneous values, it is necessary to harmonize the data through a screening before their storage into the database. The database is a PostgreSQL instance extended with spatial and geographic features, PostGIS.[7] 451Research[8] estimates that around 30% of tech companies use PostgreSQL for core applications as of 2012. PostgreSQL is an Open Source solution that in strong competition with proprietary database engines and is supported by a consistent community

---

[7]http://www.postgis.net.

[8]https://451research.com/.

of users. The use of the database in the system is twofold: first, to store the data collected by the AVL. An entity-relationship diagram defines how data is scattered among the tables. Then, the database is also needed for the administrative activities required to manage the portal. An instance of JasperServer is responsible for dashboards creation. JasperServer is a Business Intelligence (BI) tool and a reporting and analytics platform. With JasperServer it is possible to create single reports or dashboards faster. The dashboards are pre-defined through the JRXML markup language and are available for integration in the system. Entando Web Portal provides the user interface.

4.2 Gis Data

There are four files that describe the public transport network: bus stops, routes, paths and segments. Each one is provided as shape file in ED50 format. They will be transformed in WGS84 and stored in the geometry format of PostgreSql/Postgis in four tables. Figures 2 and 3 are examples of GIS-based representations of the bus stops and routes considered in this study, after the aforementioned transformation. The bus stops are represented by black circles, while bus routes by polylines.

To improve the performance of the web applications and data querying the 'avl-busstops' table is added into the database. This table with similar structure of the 'avl' table contains geometries in wgs84 and Simple Mercator coordinate systems. The first bus stop of each path is associated with a line geometry having the same 'from' and 'to' node.

When the map report has to be created the regularity measures will be linked to the geometries in the *avlbusstops* table then transformed in the JSON format and

Fig. 2 Bus stops localization

Fig. 3 Bus routes considered

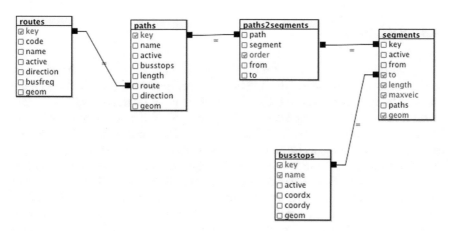

Fig. 4 GIS data schema

drawed with the appropriate color and icons. Figure 4 shows the Sql view of Gis data tables used to generate the map layers.

In the future the gis data will be published with a Geoserver Web Map Service (geoserver.org). The service is already active and provides data in GeoJSON format.[9] Overlapping the shape files we obtain the complete geo representation of the transport network (see Fig. 5).

---

[9]https://www.bi.crs4.it/geoserver/ctm/ows?service=WFS\&version=1.0.0\&request=
GetFeature\&typeName=ctm:fermatetratto\&maxFeatures=500\&outputFormat=json.

Fig. 5 Gis layer map overlay

4.3 AVL Data

In the transit industry, AVL technology helps tracking the position of the vehicles, even if most advanced systems track vehicles against their schedule, thus determining the schedule deviation and whether a bus is off-route. AVL technology was mainly developed for real time application, thus failing to capture and/or to archive data items that could be valuable for off-line analysis. The primary motivation was that transit operator did not insist on the need to archive data, thus neglecting potential benefits. Conversely, in recent years, AVL technology is adopted both for on-line and off-line applications, thus opening the doors to new opportunities. Two type of data can be collected by AVL technologies: location-at-time data, which represents the location of the bus at the time at which is polled by a center and, time-at-location data, which represents the time at which a bus passes a point of interest such as a stop, based on which schedule adherence or running time analysis is performed. To our knowledge, the first ones are mainly adopted for real time purposes, even if they present the potential to be used for off-line analysis (e.g. interpolating polling data between pairs of checkpoints), whereas the latter are more of interest for off-line analysis, even if they are usually adopted for on-line analysis as well. Since, in this chapter we look at off-line analysis, we focus only on time-at-location AVL data.

When buses run on the route, they pass at bus stops, then automatically collect time-at-location AVL data. Recorded AVL data used in this implementation are: day (data), id_route_code, id_vehicle_block, id_trip, id_bus_stop, transit_actual_time, transits_scheduled_time, dwell time. Data are stored on-board and are available on a daily basis over the entire bus network. Obviously, all buses need to be equipped with AVL technology. As vehicles end their shifts, they move back to the depot, where all data recorded during the day are downloaded through a wireless connection

```
2013-03-01 00:00:00;Z-04-1;LY0033;17:20:17;17:25:45;142;-63;2209;00ZA15
2013-03-10 00:00:00;QS-04-9;VB0990;16:15:11;16:23:53;18;-533;2057;50R01
2013-03-10 00:00:00;QS-04-9;TS0530;16:00:37;16:03:00;15;-160;1595;50R01
2013-03-10 00:00:00;QS-04-9;TS0528;16:00:01;16:01:57;15;-131;1966;50R01
2013-03-10 00:00:00;QS-04-9;TS0526;15:57:59;15:59:43;25;-126;1987;50R01
2013-03-10 00:00:00;QS-04-9;TS0506;15:58:50;16:00:29;0;-110;1985;50R01
2013-03-10 00:00:00;QS-04-9;R00498;15:53:49;15:55:34;15;-145;2134;50R01
2013-03-10 00:00:00;QS-04-9;MC0555;16:08:00;16:11:34;15;-282;1626;50R01
2013-03-10 00:00:00;QS-04-9;MC0242;16:08:38;16:13:00;15;-316;1962;50R01
2013-03-10 00:00:00;QS-04-9;MC0240;16:09:26;16:14:23;0;-342;1982;50R01
2013-03-10 00:00:00;QS-04-9;MA0493;15:56:04;15:57:45;15;-114;2084;50R01
2013-03-10 00:00:00;QS-04-9;MA0492;15:55:20;15:57:03;15;-114;2084;50R01
2013-03-10 00:00:00;QS-04-9;MA0464;15:47:54;15:51:02;19;-220;2169;50R01
2013-03-10 00:00:00;QS-04-9;LP0948;15:34:22;15:33:36;18;29;2201;50R01
2013-03-10 00:00:00;QS-04-9;LP0381;15:38:39;15:39:29;15;-65;1997;50R01
2013-03-10 00:00:00;QS-04-9;LP0379;15:37:42;15:38:10;27;-6;2198;50R01
2013-03-10 00:00:00;QS-04-9;LP0377;15:36:01;15:35:53;15;16;2200;50R01
2013-03-10 00:00:00;QS-04-9;LP0376;15:35:49;15:35:02;0;0;2188;50R01
2013-03-10 00:00:00;QS-04-9;LP0374;15:34:50;15:34:26;32;17;1974;50R01
2013-03-10 00:00:00;QS-04-9;LP0356;15:34:09;15:33:07;29;29;2201;50R01
2013-03-10 00:00:00;QS-04-9;LP0354;15:32:52;15:32:02;28;26;2110;50R01
2013-03-10 00:00:00;QS-04-9;LP0352;15:32:04;15:31:19;15;27;2207;50R01
2013-03-10 00:00:00;QS-04-9;IT0824;16:06:48;16:10:00;29;-227;1942;50R01
2013-03-10 00:00:00;OS-04-9;IT0551;16:05:49;16:09:19;25;-221;1768;50R01
```

Fig. 6 CSV file of AVL data

and stored in a central database. Next, for this application, AVL data are firstly downloaded from the central database as a CSV file (see Fig. 6), then uploaded in the our application. After that, they are sorted according to the route at hand. Finally, they are linked to other data type, as discussed in the next paragraph and processed according to the method presented in Sect. 3.

4.4 Other Relevant Data

AVL are not the only relevant data which are used to implement the method. In fact, there are two more different data types in this application: data on the Original Schedule (OS) and data on IOS. These data are matched with AVL data both to re-build the published timetables and to label which transit bus times are missed due to IOSs, that should be penalized from a user viewpoint. Indeed, as discussed in Sect. 3, OS and IOSs data are used to detect where missing data points are. Main data used to the OS are: id_route_code, id_Bus_Stop, day (data), transits_scheduled_time, dwell_time and bus_stop_order. Data used to IOSs are those reported at STEP 5 in Sect. 3.2.

Data on the OS data are first downloaded from the database of the scheduled service as a XLS file, then uploaded in our application. Next, they are separated according to routes at hand and linked to AVL data. Before using, IOSs data are manually collected by a transit operator and stored as XLS file, even if they are available in aggregate manner—i.e., per single bus trip and not per single bus transit. Hence, before being able to use them, it is necessary to carry a detailed segmentation. However, since the amount of such data is very limited and since is requires much manual work, we did not implement this type of data, even though our application allows to integrate these IOSs data.

4.5   Implementation of the Method

The implementation of the method consists in four modules, that manage the four stages of the data flow process: data import; data processing; data pre-aggregation and data management.

All the modules use PostgreSQL functions in SQL language to perform database's tasks and Entando modules in Java language to start and supervise the execution of each task as the Web application. The data are collected in a single database. A database schema is created for each month in which data are elaborated.

The main entity is represented as a table named 'AVLBusStop' described in Fig. 4. A geometric attribute contains the polyline, which starts at the previous bus stop and ends at the considered bus stop. The first bus stop of each path is associated with a polyline geometry having the same from and to node. This way, using spatial aggregate functions, one is able to merge polyline of consecutive bus stops per path code, derive spatial characteristics (geometry) of the Path entity.

The first module contains the functions that implement the first phase of the methodology illustrated in Sect. 3 the raw data of a month are imported and validated. The system loads these primary data in two tables: the 'AVL' table where each row contains real transit at bus stops and the 'Scheduler' table which contains the scheduled transit. Then the system validates the 'AVL' by applying the three steps described in Sect. 3.1. The parameters of transits percentage (80%), the chi-square test value ($\alpha = 0, 05$) and the threshold of the percentage ratio between the number of bus stops that pass the chi-square test and the total number of bus stops (60%) can be changed by the analyst.

The second module contains the tools that implement the second phase of the methodology. The module generates the 'Differences' facts table. This table contains the difference between actual and scheduled headways between two consecutive buses as measure and two multi level dimensions: year, month, day and time slot are the temporal dimension; route, path and bus stop are the logical dimension.

The third module implements the third phase of the methodology. It generates the 'Regularity' facts table that contains the regularity measures evaluated over three distinct type of day aggregations: by week (4 measures), by day of week (7 measures) and by the entire month. The pre-aggregation uses the Eq. (1) to calculate the C_{vh}

Fig. 7 PL/Schema
data201303

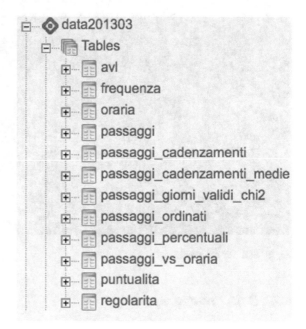

measure over each set of samples defined in Sect. 3.3. The fourth module contains a set of functions to manage data and reports. The System can also manage the JasperServer configuration for its connection to the database and regularity reports definition via the Entando/JasperServer Connector. Currently, a Mondrian Olap Cube for 'regularity' or 'differences' facts tables is not yet configured.

4.6 Data Processing

Data concerning AVL and scheduled bus transits are processed to implement the whole procedure related to the different modules. All modules are implemented using PostgreSQL functions in SQL language to perform databases tasks. Data are separated by month in a dedicated database schema. This because the focus of the methodology is related to a whole month. The public schema contains common data, such as routes, bus stops, geometries. Tables representing routes store the location of bus stops as a point.

The schema 'data201303' contains data related to march 2013 (see Fig. 7). Each step of the method is associated to specific tables.

SQL procedures are created as sql text files and executed. For instance, the procedure for the chi-square test is represented in SQL as shown in Fig. 8.

```
sql_passaggi_giorni_validi.sql ×
 1  CREATE TABLE passaggi_giorni_validi_chi2 as
 2  SELECT a.data, a.linea, a.percorso
 3  FROM chi_quadro_distribuzione b,
 4      (
 5          SELECT a.data, a.linea, a.percorso, count(a.chi_quadro_part) AS degree,
 6                 sum(a.chi_quadro_part) AS deviazione
 7          FROM
 8          ( SELECT a.data, a.linea, a.percorso, a.fermata, a.n_reali,
 9                   a.n_teorici, a.percentuale,
10                  CASE
11              WHEN a.percentuale >= 0.8::double precision
12              THEN power((a.n_reali-a.n_teorici),2)::double precision /a.n_teorici
13                  ELSE NULL::double precision
14                  END AS chi_quadro_part
15                  FROM passaggi_percentuali a ) a
16          WHERE a.chi_quadro_part IS NOT NULL
17          GROUP BY a.data, a.linea, a.percorso
18      ) a
19  WHERE a. deviazione <= b."level0.05" AND a.degree = b.grado;
```

Fig. 8 SQL for chi-square test

4.7 Data Pre-aggregation

The comprehensive procedure generates a global table containing pre-aggreated data, ready to be used by diagnostic tools. It is the 'Regularity' facts table that contains the regularity measures evaluated over three distinct type of day aggregations: by week (4 measures), by day of week (7 measures) and by the entire month. The pre-aggregation uses the Eq. (1) to calculate the C_{vh} measure over each set of samples defined in Sect. 3.3.

4.8 Data Management

The diagnostic tools are built by using some JasperServer functionalities. First, the database containing the facts tables is connected to the JasperServer business intelligence engine. Three types of reports are created through the JasperServer tools: the regularity table, the multi-line charts and the mapper. The main tool is iReport Designer, an open source authoring tool that can create complex reports through the JasperReports library. It is written in Java and is distributed with source code according to the GNU General Public License. Reports are defined in an XML file format, called JRXML, which can be hand-coded, generated, or designed using iReport Designer tool (see Fig. 9). The XML file is compiled at runtime. Designing a report means creating some sort of template, such as a form where we leave blank space that can be filled with data. Some portions of a page defined in this way are reused, others stretched to fit the content, and so on. The file contains all the basic information about the report layout, including complex formulas to perform calculations, an optional query to retrieve data out of a data source, and other functionality.

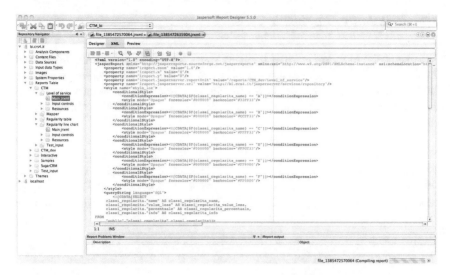

Fig. 9 File JRXML

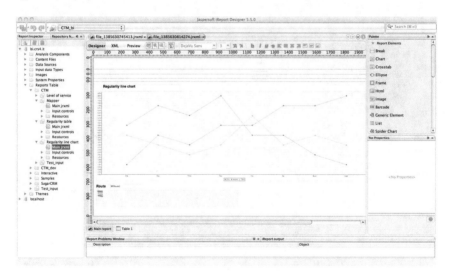

Fig. 10 Designer

To better organize the elements in the designer window, a comprehensive set of tools is provided. In the designer window, shown in Fig. 10, one can drag an element from one band to another band. In this specific case, the chart element has been selected from the palette. iReport Designer allows one to move an element anywhere in the report.

4.9 Reliability Diagnostic Tools

The diagnostic tools are realized by using some JasperServer functionalities. First, the database containing the facts tables is connected to the JasperServer business intelligence engine. Then, as shown in Fig. 11, three type of reports are created through the JasperServer tools: the regularity table, the multi-line charts and the mapper.

Each report shows the regularity measures of all bus stops of a particular route direction. For the sake of clarity in representation, the value of measures is represented by colours depending on LoS. Red colour represents LoS **F** ($C_{vh} > 0, 75$, i.e. most vehicles bunched), orange colour indicates LoS **E** ($0.53 < C_{vh} < 0.74$, i.e., frequent bunching), yellow colour shows LoS **D** ($0.40 < C_{vh} < 0.52$, i.e., irregular headways,

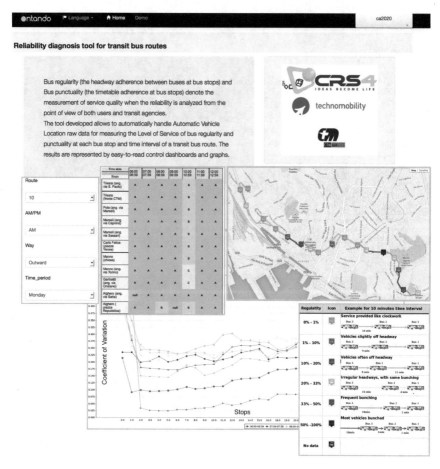

Fig. 11 Diagnostic tools

with some bunching). Other colour gradations mean LoS from **A** to **C** ($C_{vh} < 0.40$, i.e., satisfactory regularity). When LoS are not available, they are denoted by **null**.

The executive selects the route direction and the day aggregation type. Moreover, when a report is shown, the time slot can also be selected.

The regularity classes (see Fig. 11) of a bus stop in a selected time slot are represented as a table in the regularity table report, through coloured polylines and icons in the mapper report or as a coloured lines in the multi line charts report. It is important to stress that Fig. 11 is the result of different screens, the colour codes do not necessarily match. In order to permit a map representation of bus stop and path, the geometries in the 'Bus_Stops' table are transformed in the WGS84 projection and GeoJSON format using Postgis functionality and linked to the reports table.

5   Conclusion

In bus transit operators the measure of regularity is a major requirement for high frequency public transport services. It is necessary to properly account for the efficient monitoring of quality of service and for the perspectives of both bus operators and users. In this chapter, we implemented a methodology to evaluate regularity starting from the data collected by AVL. We suggested the integration of different technologies in a web portal as an environment designed to support bus transit operators experts in evaluating regularity issues. This chapter shows that it is possible to handle huge AVL data sets for measuring bus route regularity and to understand whether a missing data point is a technical failure or an incorrect operation in the service, providing a detailed characterization of bus route regularity at all bus stops and time periods by AVL technologies. The web portal ensures tool access from everywhere and anywhere. This procedure results in significant time and energy savings in the investigation of large data sets.

The next step will be to extend the web portal to both operators and users, then subsequently, to transit agencies and passengers. Illustrating the practical effectiveness of this procedure, it will be important to present a real case study. User-friendly control dashboards help to perform an empirical diagnosis of performance of bus route regularity. Transit managers can use easily-understood representation and control rooms operators following buses in real time, to focus on where and when low regularities are expected to occur. Moreover, possible cause of low level of service will be investigated, in order to put the bus operator in a position to select the most appropriate strategies to improve regularity. In addition, the method and the integration of technologies will be adapted for the measurement of punctuality in low-level frequency services.

References

1. European Committee for Standardization, B.: Transportation logistics and services. european standard en 13816: Public passenger transport service quality definition, targeting and measurement (2002)
2. Ceder, A.: Public Transit Planning and Operation: Theory, Modelling, and Practice. Butterworth-Heinemann, Oxford, England (2007)
3. Marguier, P., Ceder, A.: Passenger waiting strategies for overlapping bus routes. Trans. Sci. **18**(3), 207–230 (1984)
4. Nakanishi, Y.: Bus performance indicators. Trans. Res. Rec. **1571**, 3–13 (1997)
5. Board, T.R.: A guidebook for developing a transit performance-measurement system. Report (Transit Cooperative Research Program). National Academy Press (2003)
6. Trompet, M., Liu, X., Graham, D.: Development of key performance indicator to compare regularity of service between urban bus operators. Trans. Res. Rec. J. Trans. Res. Board **2216**(1) (12 2011), 33–41 (2011)
7. Barabino, B., Di Francesco, M., Mozzoni, S.: Regularity diagnosis by automatic vehicle location raw data. Public Trans. **4**(3), 187–208 (2013)
8. Barabino, B., Di Francesco, M., Mozzoni, S.: Regularity analysis on bus networks and route directions by automatic vehicle location raw data. IET Intell. Trans. Syst. **7**(4), 473–480 (2013)
9. Barabino, B., Casari, C., Demontis, R., Lai, C., Mozzoni, S., Pintus, A., Tilocca, P.: A web portal for reliability diagnosis of bus regularity. In Lai, C., Semeraro, G., Giuliani, A. (eds.) DART@AI*IA. Volume 1109 of CEUR Workshop Proceedings, CEUR-WS.org, 73–84 (2013)
10. Board, T.R.: Transit capacity and quality of service manual, 2nd ed. Transit Cooperative Research Program Report 100. Transportation Research Board (2003)
11. Kimpel, T.: Time Point-level Analysis of Transit Service Reliability and Passenger Demand. Portland State University (2001)
12. Mandelzys, M., Hellinga, B.: Identifying causes of performance issues in bus schedule adherence with automatic vehicle location and passenger count data. Trans. Res. Rec. J. Trans. Res Board **2143**(1) (12 2010), 9–15 (2010)
13. McLeod, F.: Estimating bus passenger waiting times from incomplete bus arrivals data. J. Oper. Res. Soc. **58**(11), 1518–1525 (2007)
14. Oluwatobi, A.: A GPS Based Automatic Vehicle Location System for Bus Transit. Academia.edu (2013)
15. Cortés, C.E., Gibson, J., Gschwender, A., Munizaga, M., Zuniga, M.: Commercial bus speed diagnosis based on GPS-monitored data. Trans. Res. Part C: Emerg. Technol. **19**(4), 695–707 (2011)

An Approach to Knowledge Formalization

Filippo Eros Pani, Maria Ilaria Lunesu, Giulio Concas
and Gavina Baralla

Abstract Just a few years after its birth, Knowledge Management drew the attention of the academic world, becoming a matter of intense study as the scientific community has always felt the need to create the cooperation among its members allowing the information exchange. Many disciplines develop several standardized formalization that domain experts can use to share information as reusable knowledge. The purpose of this chapter is the study of a process to formalize knowledge by using an iterative approach mixing a top-down and bottom up analysis of a specific domain and the bottom-up analysis for the information in the object of the specific domain. Our case study analyzes the domain of descriptions and reviews of Italian wines. We start by analyzing the information collected from the Web in order to define a taxonomy able to represent the knowledge related to the above mentioned domain.

Keywords Knowledge Management · Multimedia content · Knowledge base
Top-down and bottom-up analysis · Taxonomy

1 Introduction

Knowledge Management (KM) is a very important topic in business and in academy research [1, 2]. There are many fields of applications for KM, including Cognitive Science, Sociology, Management Science, Information Science, Knowledge Engineering, Artificial Intelligence and Economics [3–5]. Many studies on different

F.E. Pani (✉) · M.I. Lunesu · G. Concas · G. Baralla
DIEE, Department of Electric and Electronic Engineering,
Agile Group University of Cagliari Piazza dArmi, 09123 Cagliari, Italy
e-mail: filippo.pani@diee.unica.it

M.I. Lunesu
e-mail: ilaria.lunesu@diee.unica.it

G. Concas
e-mail: concas@diee.unica.it

G. Baralla
e-mail: gavina.baralla@diee.unica.it

© Springer International Publishing AG 2018   89
C. Lai et al. (eds.), *Emerging Ideas on Information Filtering and Retrieval*,
Studies in Computational Intelligence 746, https://doi.org/10.1007/978-3-319-68392-8_6

aspects of KM have been published, becoming common in the early 1990s [6–8]. Knowledge can be seen, from an operational point of view, as a valid certainty which improves the abilities of a man to undertake efficient actions. In the Information Technology context a common definition is adopted more frequently. Knowledge is considered as the information stored in human minds, thus an interpreted and subjective information concerning facts, procedures, concepts. Knowledge is not radically different from information but, as for this definition, we could say that information becomes knowledge when it is processed by a person's mind. The purpose of KM is, therefore, to keep at the whole company's disposal all the competences acquired by each of its member, so that knowledge becomes a shared, usable and protected over time *asset*. An opportunely managed knowledge can be used to easily find answers to problems already dealt by other employees, and for which the company has already invested resources, as well as to draw information to be addressed to new member's education [2]. Just a few years after its birth, KM drew the attention of the academic world, becoming a matter of intense study. The scientific community has always felt the need to create the cooperation among its members allowing the information exchange. From our point of view KM can be initially defined as the process of applying a systematic approach to capture, structure and manage knowledge, and to make it available for sharing and reuse [9]. Many approaches to information tend to use sophisticated search engines to retrieve the content. KM solutions have demonstrated to be the most successful in capturing, storing, and consequently making available the knowledge that has been rendered explicit, particularly in learned lessons and best practices. The purpose of this chapter is the study of a process in order to identify and locate knowledge and its sources within the domain, paying attention to multimedia objects. Valuable knowledge is then translated into explicit form through formalization and codification of knowledge, in order to facilitate its availability. We formalize knowledge by using a mixed-iterative approach, where top-down and bottom-up analysis of the domain to represent. We start from the concept of the domain knowledge base. The fundamental body of available knowledge on a domain is the knowledge valuable for its users. We need to represent and manage this knowledge, in order to define an ad hoc formalization and codification. After this formalization we can manage this knowledge by using semantic repositories. The chapter is structured as follows: in the second section we present the state of the art and in the third section we present our proposed approach. Section 4 is concerned with the case study, the analysis of results and verification. The last section includes the conclusion and a discussion on future work.

2   State of the Art

In recent years, the development of models to formalize the knowledge has been studied and analyzed. The ontologies—explicit formal specifications of the terms in the domain and relations among them [10]—take an important part in these formalization approaches. Ontologies have become common on the World Wide Web at the end of 2000. In the Web range there use many directory services of Web sites: the most famous is Yahoo. These directory services are large taxonomies which organize Web sites in categories. Other systems categorize products for e-commerce purpose: the most famous is Amazon. They use an implicit taxonomy to organize the products for sale by type and features. The World Wide Web Consortium W3C[1] developed the Resource Description Framework (RDF) [11, 12], a language for encoding knowledge on the Web pages. The main purpose of RDF is to make data understandable to electronic agents focused on searching for information, which is the main foreground concept of the Semantic Web. The Defense Advanced Research Projects Agency (DARPA)[2], in cooperation with the W3C, developed DARPA Agent Markup Language (DAML) by extending RDF with a more expressive constructs aimed at facilitating agent interaction on the Web [13]. The Web becomes clever and is conceived as a big database in which data are orderly classified. *Information*, therefore, is one of the keywords at the base of the success of both search engines (Google,[3] Yahoo,[4] Bing,[5] etc.), which become more refined in data retrieval, and social networks (YouTube[6], Facebook[7], Twitter[8], Flickr[9], etc.), which allow exchange and sharing, creating an interconnection among users and content makers. However, such data, despite being formally available, are often unreachable as for their semantic meaning and cannot be used as real knowledge. Various proposals to solve these problems can be found in literature, also to overcome the semantic heterogeneity problem [14] and to facilitate knowledge sharing and reuse [15, 16]. In [17] an approach ontology based approach to make annotating photos and searching for specific images is described. In [18] authors proposed a data-driven approach to investigate semi-automatic construction of multimedia ontologies. With the emergence of the Semantic Web, a shared vocabulary is necessary to annotate the vast collection of heterogeneous media; infact [19] an ontology to provide a meaningful set of relationships which may enable this process is proposed. Particularly, in [20, 21] the managing and representing knowledge problem which can be found on the Internet is discussed. Briefly, the authors classified the User Generated Content (UGC), by using a top-down (TD) and bottom-up (BU) combined approach. To reach such target, they built an ontology to define a repository of multimedia contents, putting a special focus on georeferenced multimedia objects. In conclusion, this ontology allowed for the construction of a repository to store the information extracted from UGC. Inspired by these works, we used the same kind of approach on the wine's domain.

---

[1] World Wide Web Consortium (W3C), http://www.w3.org.

[2] Defense Advanced Research Projects Agency (DARPA), http://www.darpa.mil.

[3] Google, https://www.google.com.

[4] Yahoo, https://www.yahoo.com.

[5] Bing, https://www.bing.com.

[6] YouTube, https://www.youtube.com.

[7] Facebook, https://www.facebook.com.

[8] Twitter, https://twitter.com.

[9] Flickr, https://www.flickr.com.

3 Proposed Approach

Many disciplines develop several standardized formalizations that domain experts can use to share information as reusable knowledge. According to Noy and McGuinness [30]: *an ontology defines a common vocabulary for researchers who need to share information in a domain. It includes machine-interpretable definitions of basic concepts in the domain and relations among them. Why would someone want to develop an ontology? Some of the reasons are*:

- *to share common understanding of the structure of information among people or software agents*;
- *to enable reuse of domain knowledge*;
- *to make domain assumptions explicit*;
- *to separate domain knowledge from the operational knowledge*;
- *to analyze domain knowledge*.

Sharing common understanding of the structure of information among people or software agents is one of the most common goals in developing ontologies [10, 22]. *For example, suppose that several different Web sites contain medical information or provide medical e-commerce services. If these Web sites share and publish the same underlying ontology of the terms they all use, then computer agents can extract and aggregate information from these different sites. The agents can use this aggregated information to answer user queries or as input data to other applications.*

Enabling reuse of domain knowledge was one of the driving forces behind our studies. Analyzing domain knowledge is possible if a formal specification of the terms and their structure is available. The basic concepts of our approach are the following:

1. there is no *correct* way or methodology to develop ontologies and in general to analyze and codify/formalize knowledge;
2. the main goal is to make the knowledge of a specific domain available and reusable for specific purpose;
3. the formalized knowledge is not all the knowledge in the domain, but only the interesting information for the specific problem.
4. not-formalized information have to be inventoried (they can be included in the multimedia objects);
5. the formalized information will be represented using metadata;
6. the structure of the knowledge of the specific domain has to be analyzed using a TD approach;
7. the specific domain information have to be analyzed using a BU approach;
8. the analysis process will be iterative mixing TD and BU approaches.

In fact, we want to represent knowledge through a mixed-iterative approach, where TD and BU analysis of the knowledge domain, are applied: TD and BU are typical approaches for this kind of problems. In this work, they are with further refinements, in order to achieve an efficient domain formalization [23].

3.1 Top-Down Phase: Description

We can refer to a schema driven or TD elaboration when our knowledge or our expectations are influenced by perception. A schema is a model formerly created by our experience. Typically, general or abstract contents are indicated as higher level, while concrete details (senses input) are indicated as lower level. The TD elaboration happens whenever a higher level concept influences the interpretation of lower level sensory data. Generally, the TD is an information process based on former knowledge or acquired mental schemas; it allows us to make more than what can be found in data. TD methodology starts, therefore, by identifying a target to reach, and then pinpoints the strategy to use in order to achieve the established goal. Our aim is to classify the information in the reference domain by starting from a formalization of the reference knowledge (ontology, taxonomy or others). The model could be, for instance, a formalization of one or more classifications of the same domain, formerly made in a logic of metadata. Therefore, the output of this phase will be a table containing all elements of knowledge which will be formalized through the definition of the reference metadata. We assume this phase as useful for the efficient formalization able to represent the chosen domains.

3.2 Bottom-Up Phase: Description

With this phase the knowledge is analysed by pinpointing, among the present information, the ones which has to represent together with a reference terminology for data description. When an interpretation emerges from data, it is called data-driven or BU elaboration. Perception is mainly data-driven, as it must precisely reflect what happens in the external world. Generally, it is better if the interpretation, coming from a system of information, is determined by what is effectively transmitted at sensory level rather than what is perceived as an expectation. Applying this concept, we analysed a set of Web sites containing the information of the domain of interest. From these Web sites, both information whose structure needed to be extrapolated and the information in them were pinpointed. Typically, reference Web sites for that information domain are selected, namely the ones which users mainly use to find related information. Primary information already emerge during the phase of Web sites analysis and gathering; during a first skimming phase basic information necessary to well describe our domain can be noticed. Then, important information are extrapolated by choosing fields or keywords which best represent the knowledge, in order to create a knowledge base (KB). In this phase, one of the limits could be the creation of the KB itself, because each Web site is likely to show a different structure and a different way of presenting the same information. Therefore, it will be necessary to pinpoint the information, defining and outlining them.

3.3 Iterations of Phases: Description

After TD and BU phases a classification is made and it has to reflect, in the most faithful way, the structure of the knowledge in itself, respecting both its contents and hierarchy. In this phase we will try to reconcile these two domain representations. Thus, we want to pinpoint, for each single TDs metadata, the knowledge of each object. We assume that this type of representation describe the knowledge we want to classify by, considering the semantic concept and not the way to represent it; in fact most of times it is stated in a subjective manner. Starting from this KB, further iterative refining can be made by re-analyzing the information in different phases:

1. with a TD approach, checking if the information can be formalized;
2. with a BU approach, analyzing if some information of the Web sites can be connected to formalize items;
3. with the iterations of phases by which these concepts are reconciled.

We only applied our approach to relevant data. The knowledge we want to represent is the only one considered interesting by users. At the end of this analysis we will define a formalization, in form of ontologies, taxonomies, metadata schema, in order to represent information. The final result of these phases will provide a formalized knowledge easy to be managed, reused and shared by using knowledge semantic repositories.

4 Case Study

In this paper we analyze as domain of interest the Italian wines descriptions and reviews available on the Web. The world of wines is rich in contents and complete enough to give a good starting point for our study. This is one of the main reasons we chose this domain. The knowledge we want to represent is gathered, in our opinion, by the most important and looked up wine Web sites. At the end of this analysis we will define a taxonomy able to represent the knowledge of interest for the chosen domain. We found some new items in the proposed classification, which did not exist in taxonomy or ontology used as reference in the TD phase. These new items emerged from the BU analysis. The final result of this phase will be a reference taxonomy, where, for each item, there is a linked information about the semantic definition and related to the knowledge of interest found on each analyzed Web site.

4.1 Related Concepts and Knowledge Base of Interest

Over the last decade a broader knowledge of the Web has strengthened and fostered the new applications development. The Web has turned into a multifunctional platform where users no longer get the information passively becoming authors and makers. This has mainly been possible thanks to the developing of new applications which allow users to add contents without knowing any programming code. The social value which the Web has acquired recently is therefore unquestionable; the Webs structure grows and changes depending on the users needs, becoming every day more complex. The new frontier for the Internet is represented by the Web 3.0 [24]. Due to the Web evolution into its semantic version, a transition to a more efficient representation of knowledge is a necessary step. Particularly, data are no longer represented just by of their structure description (syntax) but also by their meaning (semantics). In fact, a data can have a different meaning depending on the contexts; the use of tools like ontologies and taxonomies helps the classification of information, as shown also in [18, 25–29]. In this work, we chose the domain of wines and, particularly, the one belonging to the technical files and/or descriptions of Italian wines. Contents on wine available on the Web are thousands, offering a significant KB. Our study takes into consideration a subdomain of wine, represented by reviews we considered the most important which can be found on the Internet. Starting from this analysis, we chose a list of suitable and representative Web sites. During the Web sites selection, we took into consideration their Google ranking, their popularity and their reliability. We list our choice as follow:

Decanter.com
DiWineTaste.com
Lavinium.com
GamberoRosso.it
Vintrospective.com
Snooth.com
Vinix.com

These Web sites are considered as representative for our study because of their own information structures, we evaluated as various and different. Each web site has its own structure and a different representation of the information. To correctly define our domain it was therefore necessary to accurately analyze the contents and the layouts in each of them. Typically, the structure of the page shows the review, this is useful to understand if within the same web site structure and fields do not change. Unfortunately, we saw that some of them show the same information in a different way depending on the review, using, for instance, different tags for the same information. This, obviously, is a limit in the contents classification process. It is thus necessary to align the different items used to represent the same information within the same website.

4.2 Top-Down Phase: Implementation

During the TD phase we analyzed the existing formalizations to represent the domain knowledge. A very interesting formalization which we pinpointed was the one by the

Associazione Italiana Sommelier (AIS, Italian Sommelier Association), it provides a detailed description of all the terms associated with wine. Another important formalization was the one by the European law defining the reference features of a certain wine, such as type, colour, grape variety, etc. From these two standards, a reference taxonomy for those features was created. As an additional formalization, we chose a reference schema. By using an ontology made by W3C: Wine ontology.[10] An ontology is surely more complex than a taxonomy. It has, apart from class hierarchies, property hierarchies with cardinality ties for the assignable values. It offers a general view of the world of wines, with a less detailed description of certain fields as stated on the reviews we found on the Web. Moreover, from this ontology we took into consideration only the areas of interest related to our classification. Starting from these reference formalizations, a first taxonomy in which we pinpointed the items to create the reference table was built. After choosing the items of interest in the reference ontology, by directly extracting from OWL code we analyzed the correspondence among tags of the two. To standardize our taxonomy we decided to take into consideration the RDF standard indicating for the items with a correspondence, its URI (Table 1).

The RDF standard allows to associate a URI also to the properties. The list of the common ones associated to their classes, which can be used to standardize our taxonomy, is shown below. Referring to Table 2, in the first column we indicated some useful ontology properties which can be used for our taxonomy are indicated; in the second column the classes of the taxonomy for each feature; in the third one the associated URI.

4.3   Bottom-Up Phase: Implementation

The BU phase required a detailed analysis of the contents come from selected Web sites, we started to pinpoint the information we considered as important; then we studied the structure of each single source, useful to understand the existing data and their position in the page layout. Once the KB for the domain of interest composed by the Web sites data was defined, the next step was to classify all the chosen information. Therefore, such classification is made by considering the BU contents, namely it was built from the bottom or from the Web sites. We start considering this type of data in order to reach a general classification. One of the initial steps of our project contemplated the study of the structure of each source, useful to see the existing data and their position on the page's layout. This procedure happens to be important because allows for the evaluation of the information's classification. Both the item *maturazione*, but also the organoleptic analysis (visual, olfactory and gustatory test) if existing, are systematically shown, within the Web sites into the area we identified as *"tasting notes"*. For this reason, to build the hierarchy we tried to follow the existing structure. The type of information was also revealed both during the data analysis

---

[10]Wine ontology, http://www.w3.org/TR/owl-guide/wine.rdf10.

Table 1 Correspondence among tags

Our taxonomy	W3C ontology	URI
wine	wine	http://www.w3.org/TR/2003/PR-owl-guide-20031209/wine#
wine.winery	wine.winery	http://www.w3.org/TR/2003/PR-owl-guide-20031209/wine# Winery
wine.color	wine.wineDescriptor.winecolor	http://www.w3.org/TR/2003/PR-owl-guide-20031209/wine# WineColor
wine.color.white	wine.wineDescriptor. winecolor.white	http://www.w3.org/TR/2003/PR-owl-guide-20031209/wine# white
wine.color.red	wine.wineDescriptor. winecolor.red	http://www.w3.org/TR/2003/PR-owl-guide-20031209/wine# red
wine.color.rose	wine.wineDescriptor. winecolor.rose	http://www.w3.org/TR/2003/PR-owl-guide-20031209/wine# rose
wine.grape	Food.grape.winegrape	http://www.w3.org/TR/2003/PR-owl-guide-20031209/wine# WineGrape
wine.state	wine.region	http://www.w3.org/TR/2003/PR-owl-guide-20031209/wine# Region
wine.tasting-Notes.vintageNotes	vintage	http://www.w3.org/TR/2003/PR-owl-guide-20031209/wine# Vintage
wine.tasting	wine.wineDescriptor.WineTaste. Wine-Flavor	http://www.w3.org/TR/2003/PR-owl-guide-20031209/wine# WineFlavor
wine.tasting-Notes.TasteAnalysis	wine.wineDescriptor.WineTaste. Wine-sugar	http://www.w3.org/TR/2003/PR-owl-guide-20031209/wine# WineSugar

Table 2 Association propertiesCorrespondence among tags

Property	Classes	URI
hasWinery	Wine-Winery	http://www.w3.org/TR/ 2003/PR-owl-guide- 20031209/wine# hasMaker
hasWineDescriptor.hasColor	Wine-Color	http://www.w3.org/TR/ 2003/PR-owl-guide- 20031209/wine# hasColor
madeFromFruit.madeFromGrape	Wine-Grape	http://www.w3.org/TR/ 2003/PR-owl-guide- 20031209/wine# madeFromGrap
locatedIn	Wine-Region	http://www.w3.org/TR/ 2003/PR-owl-guide- 20031209/wine# locatedIn
hasVintageYear	TastingNotes-VintageNotes	http://www.w3.org/TR/ 2003/PR-owl-guide- 20031209/wine# hasVintageYear
hasWineDescriptor.hasFlavor	TastingNotes-WineFlavor	http://www.w3.org/TR/ 2003/PR-owl-guide- 20031209/wine# WineFlavor
hasWineDescriptor.hasSugar	TastingNotes-WineSugar	http://www.w3.org/TR/ 2003/PR-owl-guide- 20031209/wine# WineSugar
hasWineDescriptor.hasBody	TastingNotes-WineBody	http://www.w3.org/TR/ 2003/PR-owl-guide- 20031209/wine# WineBody

phase and during the study of semantics and standardization. With the creation of the tables we tried to represent the knowledge in the shape of fields as faithful as possible to those ones already existing in the samples taken into consideration. The evaluation of this phase is subjective and left to the intuition of the analyst, which freely interprets the information and intuitively obtains the taxonomic tree. This step happens to be very tricky, because is susceptible to accidental mistakes. However, we could say that the various structures found during the analysis, apart from the caption used to define each field, are not so different, thus the classification did not raise any big doubt. We built a macrosystem made of 7 tables, one for each Web site containing their information.

4.4   Iteration of Phases: Implementation

During this phase, the fields existing in the TD taxonomy (defined in the TD phase) were compared to the fields defined in the Bu phase (within the tables created for each web site). To do this, we built a macro-table mapping knowledge containing, for each item of the taxonomy, the corresponding information found during the BU phase with some additional details. Basically, within different column we indicated if and were the information exists in the Web sites but also which were not represented by the taxonomy.

In addition, to each item we assigned a numerical value to better explain the mapping. We list below the used rules:

1. existing and extractable information;
2. existing but not extractable information;
3. sometimes existing and extractable information;
4. sometimes existing but not extractable information;
5. always missing information.

For the fields with values 1 and 3, the corresponding field and the mapping rule to extrapolate the information are indicated. The information with value 2 and 4 is embedded (hidden in the text) and, therefore, should be specifically extracted by using tools of semantic analysis. Anyway, we indicated the reference fields in Table 3.

With this analysis and classification of every single data, we managed to solve the heterogeneity of the information existing in the Web. This allowed to study both its structure and the type of existing information, giving us the chance to examine how data are presented and how the classification is shown for each Web site. When creating the taxonomy, which wants to be a semantic classification tool, we also tried to represent the structure of data and the existing hierarchies of the sample Web sites. This activity was iteratively repeated to best represent the knowledge and its described connections in the macro-system above mentioned. As expected, not all the fields were taken into consideration, neither among those existing in the initial taxonomy nor among the extrapolated ones. The fields which appear just once in the whole macro-system were rejected (evaluation made considering the field value = 5), such as, for instance, *"Bicchiere consigliato"* or *"Temperatura di servizio consigliata"*. The inhomogeneity among the information coming from different Web sites was analyzed by looking for the semantic correspondences represented in the macrosystem within a column *"field details"*. The same principle was used to uniform fields with numerical values. The final range takes into account the classification used by the majority of Web sites. A simplifying table summarizing the procedure of classification above described is shown (Table 4). The result of these phases was the KB formalized through the TD taxonomy. Table 4 shows some items of it, having a mapping value 1 or 3, namely information expressed in textual form (for instance, those ones directly extractable through tags or metadata). Other fields, represented by an icon, were rejected, though their presence was considered.

Table 3 Classification

Macrosystem items	Final tags
Wine's identification	name wine
\<Produttore\> \<Winery-Producer\>	Winery: address, telephone, fax, e-mail, web, map, other wine, other info winery
\< classification\> \< denominazione\> \< tipologia\>	Classification: Vino da tavola, IGT, DOC, DOCG
\< tipologia\> \< type\>	Colour: white, rose, red
\< type\> \< tipologia\>	Specification
Qualification: embedded	Qualification: classic, reserve, superior
\< typical grape composition\> \< Varietal\> \< vitigni\> \<uve\>	Grape
\< titolo alcolometrico\> \< alcohol\> \< alcol\>	Alcohol
Label	Label
\< origin \> \< region\>	
\<origin\> \<region\> \<zona\>	State/Region
\<tasting notes\> \<reviews\> \<overview\>	Tasting notes
\<prezzo enoteca\> \<prezzo\> \<starting at\> \<$\> \<average bottle price\>	Price
\<abbinamento\> \<suggested recipe pairings\> \<food pairing suggestions\>	Food pairing suggestion
\<posted by\> \<source\>	Author
\< posted on\> \<inserito\> \<degustazione in data\>	Date
\<decanter rating\>: max 5 stelle \<rated\>: max 5 bicchieri \<valutazione\>: max 5 chiocciole \<punteggio\>: max 5 diamanti \<voto\>: max 5 chiocciole- Punteggio: max 3 bicchieri	Rate: 60–70; 71–75; 76–80; 81–85; 86–90; 91–100

4.5 Analysis of Results and Verification

During the validation phase we verified how our KMS made the acquired knowledge usable for systems compliant with others wines' ontologies and for other Italian wines reviews web sites. We verified how the chosen Web sites contents could be represented and managed on the KMS by using some simple mapping rules. Then, we tried to solve the clear inhomogeneity by paying more attention to the semantic meaning and not to the notation used to represent those contents. In fact, the purpose of the study was not to describe the whole world of wines, but just the part represented by the information gathered from the web. After matching the two systems, ontology and taxonomy, the information were generalized and made coherent. This allowed us to verify that our system is able to represent and combine specific information, and at the same time to understand the main variances between the two. This kind of study can also be used to enrich an already existing ontology with fields coming from a general classification. We evaluated a possible integration of such information without damaging the existing hierarchy. In this way we can have a broader and more accurate view over the analyzed domain. To continue with the verification phase of the created taxonomy, we decided to take into consideration another set of wine reviews sample gathered from the Web. The choice of the Web sites for the testing phase followed the same criteria used during the domain analysis. The main obstacle we found was due to the popularity of the product and the large amount of followers who have a very subjective way of representing the information about wine and the acquired knowledge. Here comes the need of pinpointing sources with clear, easily extractable and objective information. One of the main features which these sources should have was the presence of differentiated fields with a single notation rather than a broad textual field. So, also in this case, all the Web sites containing a large quantity of information in just a macro-textual area were rejected. In fact, these kind of Web sites, were not suitable for the testing phase. The embedded information, though fostering the acquisition of a general knowledge, do not facilitate its own structured classification. Similarly, some apparently suitable sources had very few contents, with a database so poor that it did not mention the most appreciated wines. After these considerations, the Web sites we decided to take into consideration for the tests were the following:

guida-vino.com
vinogusto.com
kenswineguide.com
buyingguide.winemag.com.

4.5.1   Testing Phase

In this testing phase, for each web site sample, we verified the match of the gathered information in our proposed classification, and whether our taxonomy could be able to represent them. For each Web site, we represented the specific fields of information common to every analyzed review in (Table 4).

Given the obtained results we saw that the classification defined in our study reflects the type of needed contents. Such classification, therefore, is usable, reusable and possibly extendible to the wine's domain.

5   Conclusion

In recent years, the development of knowledge formalization models has been studied and our proposed approach is a simple process of applying a systematic analysis to capture, organize and manage knowledge. Our real goal is to make interesting knowledge available for sharing and reuse. We focus our attention on interesting information that has to be represented. We applied this approach to the KB of Italian

Table 4 Classification

Existing information	Field details	Taxonomy item
Label	Label's image	Wine.label
Producer	About the producer	Wine.winery
Classification	IGT, DOC, DOCG	Wine.classification
Grape variety	Grape variety	Wine.grape
Range of prices	Price	Wine.Price
Other years	Other years	Wine.winery.infoWinery. otherWines
Presentation/comments	Wine tasting	Wine.tastingNotes
Rate: max 5 stars	Rate	Wine.rate

wine reviews collected from the analysis of many Web sites, above all we chosed a list of some suitable and representative ones after considering their popularity and reliability. We defined a taxonomy able to represent knowledge by using a mixed iterative approach, based on a top-down and a bottom-up analysis to define a reference taxonomy. These contents are to be classified in the before mentioned also using ad hoc mapping rules. The created taxonomy allowed for a definition of the reference knowledge which could be managed as an actual usable knowledge, fostered by all information existing on the selected Web sites. We chosed to validate the resulting taxonomy by verifying how the KMS allowed to make the acquired knowledge usable and accessible to the systems compliant with the Ontology of Wines. We validated the taxonomy by analyzing the content that come from Italian wine reviews web sites, underlining how, also in this case, the collected information could be represented and managed by using some simple mapping rules. A further interesting development could be the creation of semantic repositories able to collect the information previously classified. Through an ad-hoc made system the knowledge would be presented to the final user in a structured and customized way. In addition we are looking for the possibility to develop graphic interfaces with the aim of drawing the users' interests.

References

1. Gashaw, K.: Knowledge management: an information science perspective. Int. J. Inf. Manag. **30**(5), 416–424 (2010)
2. Jakubik, M.: Exploring the knowledge landscape: four emerging views of knowledge. J. Knowl. Manag. **11**(4), 6–19 (2007)
3. Dalkir, K.: Knowledge Management in Theory and Practice. Elsevier Butterworth-Heine-mann (2005)
4. Rowley, J.: The wisdom hierarchy: representations of the DIKW hierarchy. J. Inf. Sci. **33**(2), 163–180 (2007)

5. Wild, R., Griggs, K.: A model of information technology opportunities for facilitating the practice of knowledge management. In: VINE (Journal of information and knowledge management systems), vol. 38(4), pp. 490–506 (2008)
6. Ajiferuke, I.: Role of information professionals in knowledge management programs: empirical evidence from Canada. Informing Sci. J. **6**, 147–157 (2003)
7. Blair, D.C.: Knowledge management: hype, hope, or help. J. Am. Soc. Inf. Sci. Technol. **53**(12), 1019–1028 (2002)
8. Chua, A.Y.K.: The dark side of knowledge management initiatives. J. Knowl. Manag. **13**(4), 32–40 (2009)
9. Nonaka, I., Takeuchi, H.: The Knowledge-creating Company: How Japanese Companies Create the Dynamics of Innovation. Oxford University Press, New York (1995)
10. Gruber, T.: A translation approach to portable ontology specification. Knowl. Acquis. **5**, 199–220 (1993)
11. Brickley, D., Guha, R.V.: Resource Description Framework (RDF) Schema Specification. Proposed Recommendation, World Wide Web Consortium (1999). http://www.w3.org/TR/PR-rdf-schema
12. Lassila, O., Swick, R.: Resource Description Framework (RDF): Model and Syntax Specification. Recommendation W3C (1999). http://www.w3.org/TR/REC-rdf-syntax
13. Hendler, J., McGuinness, D.L.: The DARPA agent markup language. IEEE Intell. Syst. **16**(6), 67–73 (2000)
14. Euzenat, J., Shvaiko, P.: Ontology Matching. Springer, Berlin (2007)
15. Fensel, D., Van Harmelen, F., Horrocks, I., McGuinness, D.L., Patel-Schneider, P.F.: OIL: an ontology infrastructure for the Semantic Web. In: Intelligent Systems, vol. 16, Issue 2, pp. 38-45. IEEE, Mar–Apr 2001
16. Gmez-Prez, A., Corcho, O.: Ontology languages for the Semantic Web. In: Intelligent Systems, vol. 17, Issue 1, pp. 54–60, 19 Jan–Feb 2002
17. Schreiber, ATh, Dubbeldam, B., Wielemaker, J., Wielinga, B.: Ontology-Based photo annotation. IEEE Intell. Syst. **16**, 66–74 (2001)
18. Jaimes, A., Smith, J.: Semi-automatic, data-driven construction of multimedia ontolo- gies. In: Proceedings of IEEE International Conference on Multimedia and Expo (ICME), vol. 2 (2003)
19. Jewell, M.O., Lawrence, K.F., Tuffield, M.M., Prugel-Bennett, A., Millard, D.E., Nixon, M.S., Schraefel, M.C., Shadbolt, N.R.: OntoMedia: an ontology for the representation of heterogeneous media. In: Multimedia Information Retrieval Workshop, ACM SIGIR (2005)
20. Lunesu, M.I., Pani, F.E., Concas, G.: An Approach to manage semantic informations from UGC. In: Proceedings of the 3rd International Conference on Knowledge Engineering and Ontology Development (KEOD 2011), Paris, France (2011). ISBN: 978-989-8425-80-5
21. Concas, G., Pani, F.E., Lunesu, M.I., Mannaro, K.: Using an ontology for multimedia content semantics. In: Lai, C., Giuliani, A., Semeraro, G. (eds.) New Challenges in Distributed Information Filtering and Retrieval, Studies in Computational Intelligence, vol. 515. Springer, Berlin (2014)
22. Musen, M.A.: Dimensions of knowledge sharing and reuse. Comput. Biomed. Res. **25**, 435–467 (1992)
23. Pani, F.E., Concas, G., Lunesu, M.I., Baralla, G.: An approach to manage the web knowledge. In: Proceedings of International Conference on Knowledge Engineering and Ontology Development (KEOD), Algarve, Portugal (2013)
24. Berners-Lee, T., Hendler, J., Lassila, O.: In: The Semantic Web, Scientific American, pp. 28–37 (2001). http://www.scientificamerican.com/2001/0501issue/0501berners-lee.html
25. Paliouras, G., Spyropoulos, C. D., Tsatsaronis, G. (eds.): Knowledge-Driven Multimedia Information Extraction and Ontology Evolution, Bridging the Semantic Gap, Lecture Notes in Computer Science, vol. 6050, 1st Edn., IX, p. 245 (2011). ISBN 978-3-642-20794-5
26. Decker, S., Melnik, S., Van Harmelen, F., Fensel, D., Klein, M., Broekstra, J., Erdmann, M., Horrocks, I.: The semantic web: the roles of XML and RDF. Internet Comput. IEEE **4**(5), 63–73 (2000)

27. Maedche, A., Staab, S.: Ontology learning for the Semantic Web. In: Intelligent Systems, vol. 16, Issue 2, pp. 72–79. IEEE, Mar–Apr 2001
28. Davies, J., Fensel, D., van Harmelen, F.: Towards the Semantic Web: Ontology-driven Knowledge Management. Wiley (2003). ISBN: 9780470848678
29. Simperi, E.: Reusing ontologies on the semantic web: a feasibility study. Data Knowl. Eng. **68**(10), 905–925 (2009)
30. Noy, N.F., McGuinness, D.L.: Ontology development 101: A guide to creating your first ontology (2001)

Printed in the United States
By Bookmasters